TESTIMONY

FRANCE, EUROPE, *and the* WORLD
in the TWENTY-FIRST CENTURY

Nicolas Sarkozy

Translated from the French by Philip H. Gordon

HARPER PERENNIAL

NEW YORK • LONDON • TORONTO • SYDNEY

This work comprises new text and the original text of *Testimony* (published in 2006 by Pantheon Books, a division of Random House, Inc.). The original work comprised original and new text from *Libre, Témoignage,* and *Ensemble,* all published in France by XO Éditions. (*Libre* copyright © 2001 by Éditions Robert Laffont/XO Éditions. *Témoignage* copyright © 2006 by XO Éditions. *Ensemble* published 2007 by XO Éditions.)

A hardcover edition of this book was published in 2007 by HarperCollins Publishers.

FIRST HARPERPERENNIAL EDITION PUBLISHED 2007.

Designed by William Ruoto

Library of Congress Cataloging-in-Publication Data is available upon request.

ISBN: 978-0-06-149824-4
ISBN-10: 0-06-149824-6

07 08 09 10 11   ID/RRD   10 9 8 7 6 5 4 3 2 1

TESTIMONY

CONTENTS

PREFACE

I am delighted that this book is being made available to readers around the world. To be honest, I hadn't really planned to do this international edition until people started approaching me about it. I would never have thought that a French presidential election would generate so much interest abroad. But with so many people around the world trying to get hold of my speeches and my writings, I started to realize that the interest was there, and that it deserved a response.

What explains such interest? I think observers outside France saw that the presidential election of 2007 would be different from those in recent years, and that France was at a turning point. After the extreme right made it into the second round of the presidential election in 2002, and after the failed referendum on the proposed European constitution in 2005, the French people's disdain for politics seemed to have reached a peak. The public's expectations were unusually high, but so was their skepticism about their leaders.

International interest in the French presidential election may also be an indication of one of the more positive aspects of globalization: the realization that our destinies are now so intertwined that we can no longer ignore what's happening in other countries.

Whatever the reasons, once I became aware of the interest abroad, I concluded that it would indeed be a good thing to update my book and make it available internationally. This newly revised edition includes chapters from the three books I have written since 2002 as well as previously unpublished material written since my election as president of France in May 2007. I hope that by reading it, people who are not French-speakers can get to know me—and my plans for France—better.

My election was seen by many abroad as a message to the rest of the world. I think we've first got to remember that the French people who voted for me didn't do so to send a message. They did it for themselves, because they believed that my program was best suited to deal with the challenges France faces today.

At the same time, given how ambitious my program was, and given that it proposed a break with the past, my election did prove that France is not the inward-looking, overcautious country that some people say it is. In this sense, the election did send a powerful message.

I always knew that the French were not afraid of change. Instead, they were yearning for it. They wanted to break with the status quo and rally to a common project. Most of all I think the French supported me because they understood that I was telling them the truth. I told them that France's problem was that we didn't work enough, because in France work had been undervalued for too long. I told them that only by working harder could we increase their buying power. I told them I believed that rewarding merit was a better approach than relying on state handouts or promoting egalitarianism. I was clear about all my plans before the elections so that I could implement all of them afterward without surprising anyone. I took the risk of telling the truth, and the French showed confidence in me.

Some people wonder whether the French are really going to support change, now that they've voted for it, but I'm convinced that when French voters put their faith in me, they knew what they were doing. They backed a plan for a clean break with the ideas, values, and conduct of the past. They were asking me to do what I said I would do—to change what I said I would change. They understand that in a world of both great opportunity and great risk, nothing would be more dangerous than to stand still.

A few months prior to my election, *The Economist* magazine compared me to Margaret Thatcher. Personally, I don't think you can compare Britain in the 1980s to France today—these are two very different situations, two countries with different mentalities and political cultures. And I think it's also safe to say that Mrs. Thatcher and I do not exactly share the same personality and style.

That said, it is true that there is a certain similarity between the crisis of confidence in the United Kingdom in the late 1970s and that of France today. In France, this crisis of confidence results from the fact that our politics have been lacking boldness for too long. We, political leaders, on the right and on the left, bear responsibility for this crisis, because for too long we were incapable of imagining a break with the stale approaches of the past.

The other thing Britain then and France now have in common is the remedy. I know of only one way to restore the confidence of an anxious country, and that is through political will. Mrs. Thatcher's success resulted more than anything from absolute determination, an unbreakable will to get things moving. Thatcher's story is one of exceptional leadership that thoroughly transformed the United Kingdom and laid the groundwork for its prosperity for decades to come.

The article in *The Economist* ended by saying that the real question was not whether France is reformable—it is—but whether there was anyone capable of reforming it. Well, I intend to be that person.

Reforming France is a massive undertaking, and some members of the French press have already started calling me the *hyperprésident* because I've been extremely active and visible from my first days in office. Frankly, I don't mind being criticized for doing too much in a country where for so long leaders have been criticized for not doing enough. People say, "You're going too fast!" In my view, however, that's a lot better than not going fast enough.

In my view, the president of the Republic is elected to govern, not to ruminate. Because the president is elected directly by the people, he has the legitimacy to act. Once he starts governing, however, he's got to account to voters for his actions, because there is no such thing as power without responsibility. We need to think differently about how the president does his job. I don't think major constitutional changes are needed, though I do believe that the constitution should limit the number of consecutive presidential terms to two, because energy spent on holding power is energy not spent getting things done. And I think that the president's power to nominate senior officials should be more closely overseen by the Parliament, to ensure that competence plays a bigger role than connections.

Anyone who thinks that the prime minister and his government no longer have enough to do, completely misunderstands the enormous job of running the government. Believe me, there's enough work to go around. We must get beyond the absurdity of competition between the president and the prime minister. I, Prime Minister François Fillon, and all members of the government are all on the same team. Our sole objective is to implement the program on which I was elected.

To put my program into action, I decided to open up the government to talented people outside my political family. This openness stems from my conviction that the president must be the president of all the French, not the leader of a single party or clan. With a majority in the Parliament, I wasn't obliged to reach out to the opposition—reaching out was a choice. The role of the president of the Republic is to bring people together, to speak for everyone, and to encourage diversity in France. I work for all the French, including those who didn't vote for me. I have no right to abandon part of France because it chose not to support my candidacy.

I was elected on a clear program, and I intend to implement this program. Those on the left or in the center who joined my government did so in full knowledge of this, and they agreed to help me implement this presidential program. I don't see why I should do without their talent and their energy just because they're not part of my political family. Let me add that this openness is indispensable particularly because my plans for France are so ambitious; you need a big majority to accomplish major reforms.

Where the Parliament is concerned, I want to reinforce its powers, and notably its oversight function. There is no such thing as true responsibility without real checks and balances.

One of my top priorities is to get the French economy moving again; and to do that we need to restore work and merit as the basic values of our society. The growth rate in France is one percentage point lower than it should be, because we have devalued work and done everything possible to discourage the French from working. I also want to reconcile the French with success and with a taste for risk, because a society that doesn't value success or encourage risk-taking is destined to decline.

My method is simple. A government must work along multiple tracks at the same time. It's a mistake to try to undertake reforms one after the other. Instead, by taking action in different areas at the same time you can boost confidence and create the psychological and economic conditions for faster growth. In the weeks that followed my election, I pushed through a number of key reforms: waiving taxes on overtime hours to encourage work and allow employees to earn more by working more; eliminating estate taxes for 95 percent of households so that people can pass on the fruits of their working lives to their children; and lowering the "tax shield" to 50 percent, so that no longer will people have to pay more than 50 percent of their income in taxes.

The path to long-term growth is well known: competitiveness, productivity, training, and investment. We must quickly lift the obstacles to growth in each of these areas. Labor market reform is critical. I want to bring labor and business together to think about a change in our labor laws that would give companies more flexibility while also reinforcing workers' security with a safety net. I have also proposed that we think about how to adapt job contracts to the realities of the labor market. I want to give the unemployed more support during their return to work, but more must then be demanded of them. For example, I am proposing that an unemployed worker no longer be able to turn down two consecutive job offers that correspond to his or her qualifications and are in the region where he or she lives.

The return to growth also means encouraging investment, especially in new technologies, and it means promoting small and medium-size businesses. For example, I've proposed the adoption of a French version of the Small Business Act, which would require a certain amount of public procurement to come from small and medium-size companies.

Finally, we've got to reform our universities if we want our educational system to remain competitive in world markets. Specifically, we must give them more resources and more autonomy.

Economic growth is at the heart of my project for France, because only economic growth will enable us to reach the goal I have set of returning to full employment in five years. By proceeding along all these tracks at once, we'll create enough dynamism, energy, innovation, and optimism to get France moving again.

I also have great ambitions for my country as an international actor. I know that some people are dubious about France's global role. They think it's a vestige of history and that France today is too small to have the influence around the world it did in decades and centuries past.

But it's a mistake to think that size or population determines the importance of France. Throughout its history, France has played a greater role in the world than that which its demography or geography would seem to imply. Today France's voice still counts; it is listened to and respected. France is more than its 550,000 square kilometers of area and its sixty-three million inhabitants. It is a unique country that invented the concept of human rights and has fought hard for freedom.

I am convinced that France still has a prominent role to play in the world. And this, moreover, is what I think we demonstrated during my first few months in office. Since my election, France has taken a number of major initiatives: the revival of the European Union with the agreement on a simplified treaty; the Darfur conference in Paris, which led to the creation of the United Nations/ African Union hybrid force and the resumption of peace talks; the liberation of the Bulgarian nurses being held in Libya; and the resumption of political dialogue in Lebanon. We've made progress in all these areas. France did not act alone—and did not start from scratch—in any of these cases. We built on the previous efforts of others. But in each case French took the initiative and played a critical role.

I want to make clear that we're not looking for a role just for the sake of having one; nor do we expect to be involved in every possible issue. We take action not to prove that we exist, but to be useful. We act when we can get things moving and help solve problems.

I do not intend to change everything about our foreign policy, in particular because foreign policy under my predecessor, Jacques Chirac, was in many ways exemplary. This is true of his resolute action in the former Yugoslavia, which contributed to peace in that region; of his refusal to involve France in the Iraq war; of the leading role that he played in the fight against global warming; and of his tireless pursuit of a dialogue between civilizations and his respect for cultural identities.

That said, I think it's been too long since we reexamined the pillars of our foreign policy, the objectives and values that guide and inspire it. The start of a new presidential term should allow us to move forward with such a process, which might lead to change.

Because France's voice is listened to and respected, we have a particular responsibility. I want us to be more direct in our diplomatic relationships. We've got to be frank with all our partners, including our historic allies like the United States or the United Kingdom. An open friendship of choice is the only kind of friendship I can imagine.

I also want to put values back at the heart of our foreign policy. France is France only when it defends universal values, human rights, and individual freedom all around the world. Because our foreign policy reflects our identity as a nation and carries our message around the world, it must be faithful to these values.

Having and incarnating values does not mean showing disdain for other cultures or denying their individual characteristics. Nor does it mean imposing our social model on others. But it does mean rejecting the relativism of those who believe that women and men can be denied their basic rights just because they belong to other cultures. It's a mistake to think that we have to choose between our interests and our values. Defending and promoting our values doesn't mean being naive do-gooders. On the contrary, defending and promoting our values contributes directly to our own security and prosperity.

To facilitate the process of rethinking French foreign policy, I have suggested that we consider establishing a national security council. I think this makes sense because I want to get experts from different areas involved in our thinking about diplomatic and security issues. This new body, which would include people of diverse backgrounds, would analyze and debate our foreign and defense policy. At the moment, we're still thinking about the precise form that this national security council might take, but I don't think it will look much like the one in the United States. Whereas the American council is a single structure that advises the U.S. president, our national security council will play a consultative role alongside the team of brilliant diplomats who are already working by my side, under the leadership of Jean-David Levitte.

This is not the place for a comprehensive presentation of all aspects of French foreign policy, which will be discussed in Chapter 11. But let me here at least mention some of my priorities in foreign policy, and describe how France has been pursuing them since my election.

EUROPE

The very day of my election, I promised the "return" of France to Europe. What I meant by that was that France, which has always been a motor for Europe, must remain faithful to its vocation as a leader in the process of building the European Union. This was particularly important at a time when many in France and in Europe seemed to think the French people's rejection of the EU constitutional treaty meant that we should stay out of the debate. On the contrary, I always felt that voting "no" in the referendum of May 2005 gave us a particular responsibility to revive the EU. In fact, in my view our "no" did not provoke the crisis in Europe; rather, the crisis in Europe—and of the European ideal—had created the conditions for the French "no."

So when I talked about France's return to Europe, I wanted to say that France would bear its share of the responsibility for putting the European project back on track. I've always believed that overcoming the European crisis requires action. Yet for two years, Europe was stalled, and no one dared do anything about it.

The first priority was institutional: we had to make the EU capable of acting with twenty-seven members. That's why early on, in February 2006, I put forward the idea of a simplified treaty that would include the main institutional elements of the constitutional treaty. This idea gained increasing support and was ultimately adopted by all members of the EU at the meeting of the European Council in Brussels in June.

This is a good example of the return of France to Europe: not only was a French idea behind Europe's revival, but at the meeting of the European Council in June, France played a key role in convincing our partners, including the most hesitant among them, to back the idea of a simplified treaty.

I don't intend to stop there. I want us to talk about how to improve the economic governance of the EU as well as resolve the question of its borders. I want all members of the EU to work together to tackle all the issues, including the most difficult ones, because that's what our citizens expect and that's how we'll reconcile them with the European project.

THE UNITED STATES

As you'll see in this book, I have always asserted my friendship with the United States, even during the worst misunderstandings between our two countries. I see no reason to apologize for this. My affinity for the United States does not mean that I have always agreed with everything Washington does. On the question of the Iraq war, I always thought that the United States was making a mistake, and it's to France's credit that we sought to caution our American friends.

To me, however, the transatlantic link is strong enough to survive our divergences, even those as serious as the dispute over Iraq. The roots of the friendship between our two countries go back to our shared history and common values. Our bond was forged in our common struggle against totalitarianism. France and the United States share universal values and a historic mission in the service of freedom.

I sought to emphasize the strength of this link at a time when too many others, both in France and in the United States, wanted to forget it. I got a lot of criticism for that, but I don't regret it, because I believe in the friendship between our two countries. Our relationship must be that of free partners, faithful and demanding friends. Our dialogue must be constant, frank, and based on mutual respect. I am always going to feel free to tell our American friends when I think they're wrong. We'll doubtless have other disagreements, but I am convinced that we can express these disagreements as part of a constructive dialogue without creating a crisis, without resentment, and especially without theatrics.

People keep asking me what transatlantic relations might be like after January 2009, but I am not waiting for the next administration to work closely with the United States and to deepen the cooperation between our countries on all the big issues. Opportunities to work together exist right now, and it's right now that we must seize them.

COMBATING TERRORISM

One area of tension across the Atlantic in recent years has been the debate about what the Americans call the "war on terror." Personally, I don't think we can talk about "war" in the traditional meaning of the word; but on such a serious matter, semantic issues are not the most important. Islamist terrorism has diverse and complicated roots, which are manipulated for political purposes. It can be supported by certain regimes and it can find refuge in failed states. It could become even more serious if terrorists were to gain access to nuclear, chemical, or biological weapons. No country is able to deal with this threat alone. Only intensive international cooperation will enable us to fight it effectively; thus we have to work closely with the countries the terrorists come from. We've been doing this for several years now, and we must continue to do so.

Like all free, democratic societies, France is a natural target for terrorists. What happened in New York, Madrid, or London could very well happen tomorrow in Paris. To think that we're less threatened than others would be more than a mistake; it would be madness. Having been minister of the interior for more than three years, I'm well placed to know that France is no less threatened than its neighbors. The terrorist threat in France is today, as it was yesterday, formidable and permanent. We must not let our guard down.

CLIMATE CHANGE

There is now a growing international consensus on the seriousness of the issue of climate change. This is great progress, compared with the situation in recent years, when people still denied the reality of global warming and human responsibility for it.

I'm delighted that attitudes have now changed. At the last G8 summit in Heiligendamm, I told President Bush how important it was for the United States to be involved in defining the response of the international community. And he has demonstrated a new consciousness about the issue, which I applaud. President Bush recognizes the role of humans in climate change and has agreed to consider seriously a 50 percent reduction in greenhouse gas emissions by 2050. He has also committed to working toward the successful conclusion by 2009 of negotiations on the efforts by the United Nations to define a "post-Kyoto" regime. The United States has a particular responsibility to fulfill, and I will continue to encourage it to do so.

Naturally, all the efforts of the industrialized countries will be in vain if they are not closely coordinated with large emerging countries such as China, India, Mexico, Brazil, and South Africa. France will play its full role in this necessary dialogue, working closely with its European partners. I applaud President Bush's initiative to bring together for the first time the world's leading emitters of carbon.

But France's responsibility also includes making sure that developing countries get the attention they deserve. I feel strongly that the post-Kyoto regime should be developed in the framework of the United Nations, because I want all nations to be able to express their concerns.

Dialogue between industrialized countries and emerging countries is absolutely essential, but it cannot and must not be expected to provide solutions that will apply to the entire world. Developing countries must be supported for at least two reasons. One is solidarity, because many of these countries will need help adapting to the consequences of climate change. The other is collective interest, because putting in place the conditions for clean growth and encouraging better forest management in the countries concerned will contribute to the health of the entire planet.

DARFUR

With regard to genocide or crimes against humanity, silence and inaction equal complicity. This was clear in the cases of Cambodia, Rwanda, and Bosnia and Herzegovina. We cannot allow such tragedies to happen again.

I have made dealing with the tragedy in Darfur a priority because I know that if we were to allow the first crime against humanity of the twenty-first century to take place without reacting, history would never forgive us.

Since my election, France has taken a number of initiatives to respond to this security and humanitarian crisis and to help move toward a political solution to the conflict. We have put in place an air bridge to provide assistance to the refugees and displaced persons now in eastern Chad. We also organized in Paris, in late June 2007, a meeting of the enlarged Darfur Contact Group. Following this meeting, the UN Security Council unanimously adopted resolution 1769, which authorized the deployment to Darfur of a UN–African Union hybrid force of 26,000 troops and police. Meanwhile, negotiations between the Khartoum government and the rebels resumed in August, in Arusha, Tanzania. These are major advances which, I hope, will help end the Darfur tragedy.

MIDDLE EAST

France and Europe must uphold all their responsibilities in the search for solutions to the crises in the Middle East. The initiatives we take must be, first and foremost, useful, and they must fit in with the efforts already under way on behalf of the international community. This is what we did regarding Lebanon, when in July we organized a meeting near Paris of all the players on the Lebanese political scene to help them resume their dialogue.

On the issue of Israel and Palestine, I believe the resolution of the conflict will come in the form of a negotiated and mutually acceptable solution based on the establishment of two viable, demo-cratic, and independent states living side by side in security and within secure and recognized borders.

We must stand by the Palestinians to help them build the state to which they so strongly aspire and to which they are entitled. But in no case can we compromise our values or give any ground in the face of terror and hatred. Israel's existence and security are not ne-gotiable, and nothing justifies terrorism. I have always believed that the international community is perfectly justified in demanding of any Palestinian government that it respect the three conditions laid down by the United States, the EU, the UN, and Russia as the Mid-dle East "quartet": recognition of Israel, renunciation of violence, and respect for past commitments. At the same time, I have on a number of occasions told our Israeli friends that pursuing a policy of *faits accomplis* seriously undermines prospects for peace.

Today, while the Palestinians are tearing themselves apart and Hamas has taken control of Gaza, the international commu-nity must fully back President Mahmoud Abbas in his fight against extremism and for the reconciliation of the Palestinian people. This is why in June 2007, together with our European partners, we again started providing assistance to the government put in place by President Abbas. The United States has done the same thing, while Israel has resumed tax and customs revenue transfers to the Palestinians.

I am very pleased about the plans for an international confer-ence on the Middle East and I can tell you that France will play its full role alongside the international community to help the peoples of this region, plagued by so many years of war and violence, fi-nally to find the path to peace and reconciliation.

IRAN

Iran's access to nuclear weapons is unacceptable. Together with our allies, we must continue to act so that the international community remains united and firm, as it was when it adopted two UN Security Council resolutions unanimously. This pressure on Tehran must be pursued if the Iranian regime does not change its behavior. It's now up to Iran to choose between sanctions and cooperation. In return, the international community must guarantee the Iranian authorities that it will keep its commitment: access to nuclear energy for civil purposes, if Tehran respects its international obligations.

But I don't confuse the Iranian people, a great people born of a great civilization, and their leaders. There's a real debate in Iran today, because the Iranians are well aware that the behavior of their president is isolating them on the international scene, and that this isolation carries a big price. The setback to President Ahmadinejad in the municipal elections of December 2006 proved that there was a real desire for change in Iranian society and that the Iranians did not recognize themselves in the hateful speeches of their president. The Iranian people, who have already suffered a great deal, aspire to more than the isolation to which the irresponsible behavior of their leaders condemns them.

AFGHANISTAN

In Afghanistan, French troops are participating in the international community's efforts to fight against the terrorist threat and help this long-suffering country get back on its feet and rebuild itself. France will fulfill its commitments and will show solidarity with its allies and with the people of Afghanistan.

IRAQ

As France has no troops on the ground, we're not well placed to define the timetable for the withdrawal of troops. It seems to me there are two main dangers to avoid: one is a precipitous withdrawal, which would lead to chaos; and the other is the absence of any prospect of withdrawal, which would risk provoking greater violence and play into the terrorists' hands. The wisest approach in my view would be to set a general withdrawal "horizon," whose details would be worked out by the Iraqi leaders and troops of contributing countries depending on the actual situation. In this way the Iraqis would be assured that the goal really is to give them back their complete sovereignty.

Beyond that, the only solution in Iraq is a political one. We need a "new deal" among Iraqis that would assure all segments of the Iraqi population equal access to the country's institutions and resources. That's how we can isolate the terrorists.

The tragedy in Iraq reminds us that democratic transitions are long and difficult. But these difficulties must never be used to justify acquiescence in the status quo. On the contrary, they require us to redouble our long-tern efforts. That is the responsibility of the international community.

RUSSIA

President Putin has a strong personality. He's determined, and, frankly, a sometimes difficult partner. But in the end all he's doing is defending his country's interests, which is the job of every leader on the planet. It's not for others to tell the Russians what their interests are, even if we may disagree with their methods and say so. I have no problem working with President Putin, for this simple reason: he and I both have a preference for straight talk and sincerity. He knows this well: for things to move forward, and to have a constructive relationship, you've got to speak frankly to one another, even if you don't agree.

Common sense suggests that this is the way to proceed. Russia is and will remain one of the major players on the international scene, an unavoidable partner for handling great global challenges such as terrorism and climate change and for handling regional crises. We have disagreements with Moscow on some of these. I'm thinking in particular about Kosovo, where the only solution is supervised independence, now inevitable—not the illusory maintenance of a fictitious Serb sovereignty that has been rejected by the people of Kosovo ever since that territory was placed under international administration. But on other critical subjects, such as Iran and North Korea, initial suspicions and misunderstandings have gradually given way to a convergent analysis of the threat, which ultimately enabled us to act with Russia in a coordinated manner, especially in New York, where the Security Council was able to agree on several rounds of sanctions.

Without ever renouncing our principles and our values, we must avoid the temptation to try to impose our vision of the world on Russia. Instead we must always make the effort to understand the Russians' point of view, even if we don't share it. In this way we'll have the best chance to overcome our disagreements, or, when that's not possible, to manage them as well as possible given our respective interests and the interest of international stability.

EUROPEAN DEFENSE

I have always thought it rather silly to pit European defense and NATO against each other. This makes no sense. We need both, because they are complementary and reinforce each other. Look at the figures: twenty-one of the twenty-six members of NATO are also members of the European Union, and twenty-one of the twenty-seven members of the EU are members of NATO. Who could possibly believe that one could be built in opposition to the other?

European defense is one of the great successes in the process of building Europe. It allows Europe to uphold a share of responsibility for international security. The success of EU operations in Bosnia, Macedonia, and the Republic of Congo demonstrates that Europeans can play a decisive role on their own continent as well in more distant countries. Together with our European partners, we must continue to reinforce European defense. And I would like to see more EU member states participate, because European defense policy must be a matter for all of us.

But I want to stress this: in no way do I see reinforcing Europeans defense as an alternative to the Atlantic Alliance—to which, I hasten to add, France is one of the main contributors. Our security also depends on a strong, united Alliance founded on shared values and common interests. French troops have been successfully deployed as part of NATO in a number of cases: in the Balkans after 1995; in Kosovo starting in 1999, first to stop Milosevic's ethnic cleansing and then to help stabilize the situation; and in Afghanistan against the Taliban since 2002. In the coming months, it will again be Kosovo that illustrates the complementarity between the EU and NATO, because both institutions will conduct operations there.

All that said, I do not think the Atlantic Alliance should become a global instrument that also gets involved in civilian activities. NATO is primarily a military organization as well as the preferred framework for transatlantic strategic partnership, for cooperation regarding security with our ally Turkey, and for security dialogue with Russia.

These, then, are some of my priorities as president of France. I hope these few pages have given you a better sense of what I am trying to accomplish and how I plan to accomplish it. In the pages that follow, I hope you'll also get a better sense of who I am—where I come from, how I see the world, and what I have learned in more than three decades of political life.

I love my country. I want it to have a future as great as its past. I believe it is capable of meeting all the challenges of the modern world. I believe it is strong enough not to be afraid to take inspiration from what has worked elsewhere. Regarding all these issues, may this book help provoke the sort of healthy debate that all our societies deserve.

INTRODUCTION

For as long as I can remember, I've wanted to make a difference. For me, words and ideas matter only to the degree that they lead to action. I've always had a real passion for breaking old habits, making the impossible possible, and finding room to maneuver. That is how and why from a very young age, I started to take on responsibilities and to wield what we tend to call power.

This passion might have led me into business, community affairs, humanitarian intervention, or who knows what else. Politics is not part of my family tradition. In fact everything should have led me in a different direction: I had no particular connections, no great fortune, no government position—and I had a foreign-sounding name that might have led some people to hide in anonymity rather than to step into the limelight.

I became a lawyer and I love this work. Most important, it gave me the reassurance and the certainty of having a job to fall back on. Without that, I would never have been able to take the risks that I have taken throughout my career. I owe to this profession the independence I've needed to remain a free man. It is so much easier to say no when you know your professional future is secure.

In any case, it's politics that has attracted all my interest and desire at least since I was fifteen years old. I did not choose to go into politics. I never said to myself, "I would like to go into politics." It just happened—naturally and irresistibly. That's why I've never really tried to explain it. I can think of no particular meeting, event, book, or article that played a big role in leading me in this direction. It was something deep inside me, and it would have been unnatural for me not to follow it. This is no doubt why—notwithstanding all the obstacles I've had to overcome, all the failures I've had to endure, all the tests I've had to pass—this passion has always kept me going.

Of course, on the mysterious path to any vocation, the years of one's youth necessarily leave their mark. Looking back, I would be wrong to suggest that my evolution had more to do with what I experienced in my youth than with who I am deep inside. But I would be just as wrong to say that nothing or nobody mattered in how my politics evolved.

The fact of being an immigrant certainly played a role. My father came from Hungary after the tragedy of Yalta, and my mother's father was a Jew from Salonika. It was no doubt easier to be the son of an immigrant in this France of the 1960s—when the country was drawing on all its forces to modernize and develop—than it is today. But we loved France. We didn't take it for granted. Throughout my childhood, perched on the shoulders of my grandfather, I was always fascinated and moved by the annual Armistice Day and Bastille Day parades. I never got tired of watching them, and it would never have occurred to me to criticize France.

If "admiration is at the source of every vocation," as the writer Michel Tournier has said, then I must also mention General Charles de Gaulle, and Gaullism in particular. Because I was too young at the time, my family didn't allow me to participate in the great demonstration supporting General de Gaulle after the massive student and worker protests of May 1968. But like thousands of other French men and women, I laid a flower under the Arc de Triomphe the day of the great man's funeral in 1970. Gaullism overcame all political and social divisions and brought millions of French people of different backgrounds and social classes together behind a "certain idea of France" and a desire to modernize and transform France. I was fascinated by this ability to break habits and traditions in leading an entire country to excellence.

Compared with any other type of activity or engagement, politics has the enormous advantage of letting you work with people, rather than against them or without them. This is of great interest but it's also very demanding. I had my first political experiences in the popular fervor of Gaullist rallies, which confirmed for me that getting involved in politics was not a mistake. I love people. I love meetings, interactions, and shared emotions. I love the idea of working together toward a common goal. I know this is the only way to advance, that otherwise we'll never get anywhere. I love to convince people. For me politics has meaning only when its objective is to give hope to millions of people.

Finally, I have been convinced since I was very young that you have to make your own future—otherwise you're condemned to accept whatever comes your way. It's no secret that I don't miss childhood. I was impatient to become an adult and to become free. This desire for independence made me determined to live the present with the energy of someone who knows that the promise of the future does not come automatically. "To build" is—along with "to love"—one of the most beautiful concepts in the French language. You build your house, your life, your family, your family's happiness, and sometimes your country's happiness. You love your family and you love your country passionately. You give them constant attention and energy. You must never stop, give up, or let your guard down. How could I ever forget this? As I now know, things can fall apart quickly, in private as well as in public life.

I have no intention of analyzing political power in this book. I don't claim to be putting forth any new theory or new intellectual framework. I simply want to tell the story of a life in which the desire to act decisively and get results plays a big role.

I would like to explain what I've tried to do, what I want to do, and—still further—what it is possible to do. France has been going through a fundamental crisis of confidence. The main characteristic of our society is the absence of hope, whereas the very goal of politics is to provide hope. I reject the notion of inevitability. I cannot stand the word, the idea, or the consequences of the concept. So many people have given up believing that tomorrow can provide hope. They've given up on social progress for their family. They've given up on a happier future for their children. Whatever energy remains in our society is used not for moving forward but for self-defense. Self-defense for everyone and against everything has become the last resort of too many French people.

To build something you need to take action, but only after taking the time to reflect. You've got to act, but according to a plan. Too many political leaders have lost their vision because they no longer have faith in their ability to change the future. They confuse vision with prophecy. They think they're being asked to predict the future, whereas they're really being asked to invent it. The role of politics is to propose a future and then to make it possible. That's what I have set out to do. It's why I believe in political will, and it is what justifies, in my mind, the desire to attain the highest responsibilities. To build and to love. This could be a pledge. For me, it's my life.

CHAPTER 1

Political Beginnings

Few political families are as affected by their history as Gaullism. Because of the experience of war and the solidarity born of the Resistance, this is understandable. In fact, however, recent generations of Gaullist leaders did not live through these events, and yet nothing has really changed. Nostalgia for the past, the love of epic stories, and the presence in the party of mythical and charismatic speakers are all part of the Gaullist legacy.

I remain, sometimes despite myself, deeply affected by this history. I have my favorite memories and it's hard to stop me from bringing them up. One such example is my first participation in a party conference of the Union of Democrats for the Republic (UDR)—the name of the Gaullist movement in the mid-1970s—which remains etched in my memory down to the smallest detail. It was in 1975, in Nice. I had arrived, like many other party organizers, on the overnight train. It was my first visit to the city that has since remained my favorite. It was hot, the sun was shining, and the young ladies of Nice looked almost perfect to my twenty-year-old eyes. My heart was beating even harder at the thought that the following day I would have the honor of getting up on the stage to give my first speech.

It would, admittedly, be a morning speech, and the amount

of time allotted for me—less than five minutes—was not exactly likely to make my appearance the highlight of the conference. But it felt that way to me! I had had a really short night. I couldn't sleep because the idea of giving my first speech kept spinning around in my head. I had written my text on both sides of a piece of paper, violating speech-making rules that I knew nothing about.

Entering the room Sunday morning, I could hardly breathe. Everything was enormous. It was all much bigger than I had imagined. I particularly remember the stage and the speaker's podium that looked like the stern of a ship, really high up. This was an era in which the choreography and decor had to be magnificent. Proximity to the people didn't seem to be the highest priority. My heart was beating as never before, yet I wouldn't have given up my spot for anything in the world. I was excited and terrified at the same time.

Jacques Chirac was at the time prime minister and leader of the Gaullist movement. He was chairing the conference. Ten minutes before it was my turn to speak, they came to tell me to be ready to go onstage. I was sitting there on a stool that I remember was wobbly—and I was already wondering if this was a bad sign. Then Jacques Chirac called to me: "Are you Sarkozy? You are speaking for five minutes, and you won't be given a minute more than that, understood?" I went along willingly, without really having understood what he had said. My final memory is the moment when, for the first time, I found myself standing behind the lectern. I was blinded by the light of the projectors and surprised by the sound of my amplified voice. Curiously, I don't remember anything about what happened later on this day that would determine the orientation of my life. It's as if all that mattered was the starting point.

After this first meeting with Jacques Chirac, I was with him for all the big battles over the following fifteen years. These included the creation of the Rally for the Republic (RPR), which replaced the UDR as the main Gaullist party in 1976; his conquest of the Paris mayor's office in 1977; the legislative elections of 1978; his first failed presidential run in 1981 (when I was head of the National

Youth Committee in favor of his candidacy); and his second failed attempt in 1988 (when I organized the main campaign meetings). Since I was too young and not yet in Parliament, I didn't play much of a role when the right was in power from 1986 to 1988, with Chirac as prime minister.

My political journey was a lot harder than people have often said and even than I have admitted. A lot of political leaders have found their vocation by working in a ministerial cabinet right after graduating from the National Administration School (ENA). It is much rarer to start as a grassroots party organizer and climb your way up, but that's the route I took. I was secretary for my constituency, then regional treasurer, then a leader of my region—I served at practically every basic level possible. It wasn't until ten years after I first started in politics that I became mayor of Neuilly, after the sudden death of my predecessor, Achille Peretti.

During all these years, Jacques Chirac and Charles Pasqua—a leading figure in the RPR—often asked me to work directly for our political movement, but I always energetically refused. This was because for as long as I can remember I saw working for a party to be like being in an intellectual prison, cut off from all freedom of choice. In this case, material dependence inevitably leads to political dependence. And I wanted to preserve my political freedom at all costs.

Thus I was never a paid employee either of the RPR or of Jacques Chirac. And that turned out well, because in 1983, the RPR didn't support me when I ran for mayor of Neuilly. In my path was none other than Charles Pasqua, already a senior Gaullist politician. I am grateful to Chirac, who didn't back me, for not having done anything to stop me from being a candidate, even though I was only twenty-eight years old. This independence also served me well a second time, in 1988, when I decided to run for Parliament against the incumbent deputy Florence d'Harcourt, to whom Chirac had already pledged the RPR's support.

When I won, I became a member of Parliament for the first time. And it was only at this time that I began to hold political responsibility alongside Chirac. I'll address my complex relationship with Jacques Chirac later in this book. But what I want to say here is that when I was at his side for all these political battles I was totally committed. This is the way I am. I have trouble imagining any sort of commitment other than total. Given Chirac's role in shaping the history of the Gaullist movement, it's only natural I have often been at his side.

I say "often," but not "always," as there have been exceptions, most notably in 1995. So much has been said about this that everyone still remembers it. Everything has been said, written, told, commented, imagined, and exaggerated to the point that the caricature has ended up being the reality, or at least the perceived reality. I want to talk about it here to give my version of events that I lived through passionately. After all, I think I have at least as much right to talk about them as a number of interested bystanders. I am under no illusion that I'll be able to convince people of my version of events. The facts are now old and the images in people's minds are very strong.

But I still want to give my version, once and for all. Those who, like me, supported Édouard Balladur rather than Jacques Chirac for president in 1995 have paid the price for having lost. The rules of democracy are clear. If you lose you're wrong, and if you win you're right. That's just the way it is. It serves no purpose to challenge this; it's the price of democracy.

At the beginning of the 1990s, everything was simple. Jacques Chirac needed Édouard Balladur to win the presidential election, and Balladur needed Chirac to become prime minister—or at least to get some other high position. I was working for and with both of them. As lawyers would say, there was no conflict of interest at the time.

Thus from 1988 until 1993, while organizing the annual meeting of the opposition along with another RPR politician, Alain Madelin, I was working relentlessly on what would become the platform of the opposition that would win the 1993 legislative elections. I owe this new direction in my political life to Édouard

Balladur, the moderate RPR politician who would become prime minister in 1993, and in whose government I would serve as minister for the budget. Before I met Balladur, my political repertoire consisted only of sharp elbows, cheap applause lines, ready-made ideas, and the deployment of my inexhaustible energy behind Jacques Chirac.

Working with Balladur, even though I was very different from him, I discovered several things: the advantages of compromise, tolerance, respect for skepticism, a profound commitment to consensus (or at least a strong aversion to conflict), and a healthy detachment from people and events. In short, I felt able to use new forces and to rise to a new level. I was grateful to Balladur for having considered me to be up to crossing this threshold. In contrast, a weakness of Jacques Chirac was always to want to place people around him in a box from which, if they were not careful, they would never reemerge.

A FIRST TASTE OF POWER

After the victory in the legislative elections of 1993, the question was whether I was going to join the government or instead take over from Alain Juppé as secretary general of the RPR. Jacques Chirac wanted me to take over the party, while Édouard Balladur wanted me to join the government. I wanted to join the government so badly that I had no trouble convincing Chirac to accept this, and I entered the government as minister for the budget.

Then it was time to act. For two years, I learned and evolved all the more because in government there is always some battle to be waged, some crisis to resolve, or some challenge to meet. I could thus use all my energy, with the only limit being my physical strength. Since not so many of us in government worked like this, I ended up taking on even greater responsibilities than the already significant ones Balladur initially gave me. I was all for it. The prime minister needed me. Our cooperation and agreement developed further each time we had a crisis to manage together.

I was enthralled, which is not to say that I was happy. In fact this was one of my first discoveries. From the outside, I had a superficial and actually rather silly idea about power. Having never had it, I had a sort of immature fascination with it. But reality quickly set in. After the first few weeks, when I have to admit that my head grew rather large—notwithstanding all the warnings I got from my friends and my wife—the troubles began.

I had to accept that power was no fun, and in fact that it was rather sad. The combination of derision and suspicion that people felt about politicians didn't do ministers any favors. You've got to fight, withstand criticism, and attack relentlessly, day and night. Moreover, being in government is probably the worst place in the world to think. If something hasn't been thought up, considered, and prepared before you take power, there is ultimately little chance it ever will be.

From this point of view, the American system is much better than the French one, because it gives the president a useful break between his victory and taking office. By contrast, French politicians exhausted by a hectic political campaign have to immediately choose their team and make the first decisions, which are by definition urgent. All of this happens in a week. The fallout comes quickly. You don't make the necessary decisions; you take shortcuts, and you don't get the serenity you need to exercise power.

Thus a lot of governmental debuts are disastrous, except for those of leaders who were wise enough to prepare gradually for power, which enables them to resist this form of drunkenness from which it is hard to recover. Édouard Balladur had such wisdom. Paradoxically, it was because everything was going so well that the problems arose. Balladur was a perfect fit for France after the suicide of former prime minister Pierre Bérégovoy, the failure of the Socialists, the economic crisis, and all the deficits. His personality reassured people, brought them together, and reduced tensions. Encouraged by the opinion polls (which of course would turn out to be disastrously wrong), he convinced himself that he would be the right man in the right place for the 1995 presidential election. Such a reaction was only human.

He faced challenges every day. And he dealt with them rather well. The French seemed to be backing him. How could anyone resist becoming a candidate under such conditions—especially when his relationship with Jacques Chirac was deteriorating rapidly?

I remember the first day of April 1993, the day on which we formed the government. Balladur wanted to show that he was his own man. Chirac wanted to weigh in on the nomination of every minister. We had to organize a reconciliation dinner that very evening—the new government's first day! This did not augur well for what was to come.

I must admit that I participated in all these events with enthusiasm, possibly even self-indulgence, and often without the sort of detachment that would have been appropriate. I was working every day, practically every minute, with a prime minister in whom I had confidence and with whom I saw eye to eye on practically everything. He was also a candidate for the presidency. How could I not support him? In all honesty, I must also say that I had no desire to leave the government. I loved what I was doing and believed in the direction in which we were moving. There, too, I had no reason to want to leave the ship. Looking back even a number of years later, I have to say I would have done the same thing all over again.

On the other hand, I would have gone about it differently. Experience and confrontations with failure lead you to see life differently—in a less brutal, more balanced, and more human way. At the time, I had too many certainties. I guess it would be fair to say that I had nothing but certainties. All too often, I allowed myself to see everything in terms of two irreducible worlds: friends and adversaries. I wanted it all, and of course I wanted it right away. I called rudeness "being frank." This wasn't all bad. It made me seem new and authentic—I wasn't the usual stonewalling politician and I certainly wasn't hypocritical. And I didn't care about my image, because I was sure that it would be shaped most of all by the result of the presidential election. If we won we'd be geniuses. If we lost, I understood the rules and knew what I'd be in for.

All this is what led me to tell Jacques Chirac as early as November 1993 that I would not support him. We were at RPR head-quarters on the rue de Lille. It was at the end of one of our political meetings. Chirac called me so that we could speak, just the two of us. Curiously, we found ourselves in the big press room. The climate was warm and relaxed, exactly as Jacques Chirac knows how to be when he wants something.

I didn't like the way he started the discussion: "By supporting Balladur like this, you're putting all your eggs in one basket." Even though it was realistic and ultimately turned out to be prescient, this cynical opening line put me off and left me no choice but to re-act in a candid, even blunt, way. This was our last real conversation until our reconciliation more than three years later.

Paradoxically, I was relieved to have been able to let him know where I stood. Things were now clear between us; I was going to support Balladur.

At the time, certain of my decision and my method, I had little regard for those who were waiting for the latest polls to announce what they would claim to be their heartfelt choice based on con-viction. Yet here again I have to admit that maybe they weren't as wrong as I thought, since not only did their ministerial careers not have to suffer, but their convenient indecision actually served their images rather well. In any case, it didn't hurt them.

This wasn't the case for me. To be sure, it was only natural for the new government, intoxicated with success and the feeling of power, to make me pay for having backed the wrong candidate. Still, I thought, not only was it counterproductive and inept to shut me out of government, but also the charge of treason they leveled at me was unfair.

It was unfair because while my choices could legitimately be questioned, my transparency should have protected me from any witch hunts. I took my risks and I took responsibility for them. This was rare enough in politics that one might at least have thought it would deserve some respect. And their shutting me out was inept because switching from Chirac to Balladur—who were both on the right, in the same party, and associated with similar ideas—should have protected me from the accusation of inconsistency that one so often hears in politics.

Every French person has the right to choose. I had worked for a long time with both men and for a number of years for the two together. According to what logic should I have been the only one with no right to choose? Here, too, I came to realize that this vision of the political situation was too rational, too polished, and too coherent. I was forgetting the element of drama that there is in presidential elections. The best outcome for the losers is just to be forgotten.

This was not my fate. I quickly became an object of caricature on *Les Guignols de l'Info*, the French version of the satirical British show *Spitting Image*, where puppets skewer politicians and other public figures. They just couldn't resist. I was a politician who loved politics and didn't hesitate to say so. So they had three good reasons to relish going after me. First, I had lost. And the *Guignols* have long shown that they love to kick a guy when he's down. Second, I couldn't do anything anymore for Canal Plus, the station on which the show appears, since I was no longer communications minister. Finally, and worst of all, I had had two successful years working with Édouard Balladur. We had to be punished. And the *Guignols*, who always buy into the conventional wisdom, meted out the punishment.

SUCCESS AND FAILURE IN FRANCE

The relationship in French society between failure and success is a curious one. Success is not really seen or accepted as a positive value. The explanations that are given for it, in whatever area, are often designed to put it into context, denounce it, and especially to make sure that it's never presented as an example. It is hardly a caricature to suggest that politicians are presented as dishonest, business leaders as greedy, and sports stars as drug addicts. Thus anything that is remotely associated with power is going to be criticized. This system reaches its culmination when the distrust also extends to judges, journalists, and police officers. Every form of authority is thus considered to be suspect, or even illegitimate.

All the hard work done by those who are eventually successful is rarely acknowledged. This attitude is explained by the French desire for egalitarianism, the fascination with leveling out, and, frankly, jealousy. Thus success is more often criticized than presented as a model. All elites, or those seen as elites, are targets. Everyone focuses on their advantages. Yet their responsibilities are rarely mentioned. Instead of mobilizing society through those who have succeeded the most, the French prefer to stoke up resentment of those who have more than others, on the assumption that they must have stolen what they have from others! By doing this, we're misunderstanding the real meaning of republican values.

The Republic is not about equality; it's about equity. The real republican innovation, starting with the Revolution, is that success should be based no longer on birth or fortune but on merit. And the equality of means to succeed has never meant equality of results. This egalitarian vision leads to a society in which the question is no longer "How can we give everyone the means to ensure his family's advancement?" but rather "How can we make sure that my neighbor doesn't have more than me?" At its worst, this means that it's better for everyone to arrive late than for one person to arrive on time. Far be it from me to question the necessity of redistribution, which is essential for balance within society, but why on earth pursue it so excessively and with such resentment of success?

After all, redistribution is possible only if there's something to redistribute! What's most surprising is that in such a climate one might expect that failure would become banal, or at least temporary. That would at least have the advantage of giving a second or a third chance to those who need one. But no! Failure is seen as embarrassing. Pity the poor guy who has gone bankrupt, or failed his exams, or has trouble getting going again after one of life's many tests. This first failure will often become an indelible stain, and in France he'll rarely be given a second chance.

This is all the more absurd in that failure, I have noticed, teaches you a lot more than success. How can you be certain to succeed the first time? Why should anyone be surprised that in these conditions the French are so disinclined to take risks and imagine their future? Because of our tax system, taking risks is not rewarded financially, and it's also frowned upon socially. This is how you allow a society to get old and die. Allowing success to be seen as a model and making failure acceptable—because it's part of life—seem to me to be more promising ways to move forward, at least for those who want to live their lives as actors and not mere spectators.

All this said, I have to admit that these general ideas were little consolation to me when I had to face the storm that followed the 1995 election. For a long time I remained stunned by the reactions and by the length of the penitence expected of me. I thought it was absurd that I had to stand aside and passively watch my natural political family lose ground. Yet I concluded that while it can be fierce, French society cannot be handled too roughly. You have to go one step at a time, accept its ways, and take your time. Only time gives actions the legitimacy they need.

CHAPTER 2

A Brush with Defeat

I'll never forget the date June 13, 1999. Nor will I forget even a minute of the evening following the elections to the European Parliament that day. I had thought that this, my first national political campaign, would turn out somewhat better. But I soon realized I would have to pay the price for the outcome. I got into this situation despite my better judgment, thanks to the circumstances created by Philippe Séguin's resignation from the leadership of the RPR in April 1999. Polls had shown all week that support was declining for the electoral list I was heading. The only question was how big the fall would be. I was in for a shock, as our worst expectations were realized. After six weeks of tough campaigning, during which I did my very best, I ended up with around 13 percent of the vote. Having campaigned on the themes of work and merit, I was well placed to testify that they do not always pay off! But it was too soon to draw conclusions or even analyze what happened. First we had to get through this forsaken election evening. I was focused on one thing only: keeping my dignity in the face of defeat. I chose to stay with my family while waiting for the election results. I wanted a final few moments of peace and sincere affection before confronting the wolf pack.

In the late afternoon, I went back to RPR headquarters, where I met several loyal friends and my wife, Cécilia, whose presence was more indispensable than ever. I asked my oldest son, Pierre, to be there as well. A fifteen-year-old, he was already passionate about politics. He had seen many of my successes, so I thought that it wouldn't hurt for him to see the other side of the coin as well. This way he would also see failure, and the consequences of failure. I hoped that he would remember it as a lesson for his own life. The first exit polls came quickly, and they were dreadful. I was still hoping to finish in the upper part of the range projected for me, around 14 to 15 percent. The RPR building was empty, as if the party workers knew what was coming. A buffet was set up on the same floor as my office. I devoured everything there. I wasn't hungry, but I felt I would need energy. Jacques Chirac called several times to ask about my mood and to tell me that whatever happened I shouldn't make any hasty decisions. I suspect he was thinking of my possible resignation. I responded with banality. Deep inside I had known for a long time what I would do. If I failed, I'd go.

The circumstances of the failure didn't matter. Nor did it matter that I was not the only one responsible for it. My mind was made up. I was heading our party list, and it was up to me to accept the consequences. I was just waiting to get the final results. Deep down, I still had trouble believing that the inevitable was really going to happen. Was this because I knew I had given it my all? Was it the misleading image of all the packed auditoriums on the campaign trail? Or was it the natural blindness of someone who throws everything into the electoral competition? Whatever it was, I continued to cling to the ridiculous hope that the actual results would prove all the pollsters wrong.

But of course that didn't happen. At 11:00 p.m., as I got back to campaign headquarters, I had to accept the facts. The television and radio people were waiting impatiently for a statement from me. It was hard to get away to draft my comments alone. Some real friends were around, but there were also others who had come just for the spectacle. They were easy to spot. They had refused to participate in my campaign, and this evening they were just watching in silence. I knew that their silence wouldn't last long. They would soon be putting all the blame for the defeat on me. At this particular moment, I couldn't have cared less. There would always be time later to think about and digest what had happened.

I planned to make a short statement, for what would be the point of a long one? Rejecting the friendly advice that was given me, I decided not to beat around the bush. I admitted defeat. I took full responsibility and I was ready to do what I had to do. Having lived through a number of crises, I knew all too well that there is no point in sending multiple messages. And it's a big mistake to refuse to accept reality. You've got to own up and face the consequences. It seemed to me the only possible strategy was personally to take the blame for our result. Or at least this was the only dignified strategy, and the one that would best keep future options open. And in the situation in which I found myself that spring evening, being dignified and being seen as such were about all I could hope for.

Once my declaration was made, all I wanted to do was go home, to flee all the agitation. I wanted a chance to recharge my batteries by getting away from the media frenzy that had followed me ever since Philippe Séguin left the presidency of the RPR, leaving me alone on the front lines. The cameras and microphones followed me onto the staircase, waiting for me to collapse from exhaustion and defeat. A few of the journalists who had covered the campaign affectionately patted me on the shoulder or warmly shook my hand. This is doubtless what some, who don't understand how rough politics can be, would call "collusion." They're wrong; it's sympathy at the most and basic humanity at the least. There is nothing sinister about it. I was equally grateful that most of their colleagues chose to avoid me. Starting the next morning, however, I would be the target of their sharp pens.

These were the things I was thinking about when I got a call from Jacques Chirac on my cell phone. He was unambiguous: "You've got to go on television. You have to make sure you don't give the impression that you're backing down. Take full responsibility!"

There was no point in protesting that I really didn't feel like making a spectacle of myself, especially among all those who thought they had won by winning around the same number of votes as I did, like Gaullist baron Charles Pasqua, far-right candidate Philippe de Villiers, and especially centrist François Bayrou. But the president was so insistent that I eventually gave in. What else could I do? I was in between two floors in our campaign building, surrounded by journalists, and on the telephone with the president of the Republic, the night of an electoral defeat. There are better circumstances to promote intelligent dialogue.

My son Pierre was still there. He felt the anger of a teenager ready to avenge what he saw as an injustice against his father. But he also felt emotion—he was practically on the verge of tears from thinking about how disappointed I must have been. It was past time for him to go to bed—he had school the next day. I spoke on the phone to his brothers and his half sisters who had remained at home, to reassure them. My evening continued, with Cécilia by my side.

It was thus only grudgingly that I went to the studios of TF1, the national television network. (If you have to go on TV, I reasoned, you might as well go for the most viewers possible.) But the trip to the studio surprised me. It was almost midnight, and the streets passed by without my even noticing. Lost in my thoughts, I wasn't paying attention to anything else. I only barely managed to take in the two pieces of good news that emerged. They didn't really change anything fundamental, but they boosted me at least for a minute. I edged out Charles Pasqua on his home turf in Hauts-de-Seine, and the voters of Neuilly overwhelmingly backed my list instead of the one led by the regional governor of Neuilly Nord. As the weeks went on, I came to appreciate how useful it is for a politician to have a solid electoral base—especially during political storms. But at the time, these bits of good news didn't manage to elicit from me anything more than a pathetic smile.

As always on election night, the TF1 studio was crowded with buffets around which lots of people—both the usual suspects and newcomers—huddled. Journalists, officials, politicians, staffers, and various schemers were there as usual, drawing conclusions from the results and following the evening's many ups and downs. I had to cross one of these groups to get to the makeup room. In other circumstances, people would have been pushing one another out of the way to get a comment from me or just to shake my hand. But not on this sorry evening. I was instead greeted with a heavy and embarrassed silence. The loser had arrived. They scrutinized me from top to bottom. They wanted to see all my wounds, both real and imagined. The crowd opened up before me, silent and vaguely reproachful. When I arrived on the platform with the expert interviewer Patrick Poivre d'Arvor, one of France's most famous television presenters, there was total silence. The representatives of the left were glum. Philippe de Villiers was there, exultant. In truth, this was a rather natural feeling for him. Pumped up by his new importance, he was giving his views about everything. I was hardly listening. All that came through were words such as "earthquake," "trauma," and "triumph." "Nothing will ever be the same," he concluded. I was smiling inside, thinking that if I had been up to it I would have said, "Actually, you'll always be the same." And this wouldn't have been a compliment. Before leaving the platform, Philippe de Villiers offered me his hand, but I pretended not to see it so that I didn't have to shake it. I don't know if this reaction was whim or spite, but it's what I felt. He had attacked me throughout the campaign and it was too soon to pretend to have forgotten.

Finally the moment came when Patrick Poivre d'Arvor turned toward me. He did it with the tact shown by those who have been tested and the coolness that comes from countless years anchoring the evening newscast on France's top channel. Looking me right in the eyes, he asked me what it felt like to have lost. I didn't see him as acting out of some sort of sick curiosity or even as one of those moving in for the kill. Rather, I took it as an invitation finally to be myself in this human way which people who don't know me rarely see, and which they doubt exists. And indeed, for once, I hadn't prepared, written, or calculated anything. My words came out following the emotions of the moment, without any advance reasoning. At this particular moment, I just didn't care about political calculations or even my appearance, which I admit has not always been the case. I just spoke the truth, without running away or apologizing, and without making any excuses. Failure was undeniable. I had done everything I could—if not to avoid it then at least to limit it. I didn't succeed. I knew the rules, and I had to pay the price.

Patrick Poivre d'Arvor didn't interrupt me. At most he just nudged me forward from time to time. He didn't seek to push where it hurt most. I'm still grateful today for this sort of modest respect. In this way, he actually got more out of me. The experience ended up being less difficult and less painful than I had imagined it would be. I was even relieved. It's often this way—I've noticed that challenges often don't turn out to be as difficult as they seem in advance. Does the power of our imagination lead us to exaggerate the threat? Or is it human nature that gives us greater power to react or rebound when we're faced with an obstacle? It's hard to know, but it really seems to work this way.

The evening I had so feared was finally over. The failure was behind me. Now I had to try to deal with the aftermath. Because failure and the consequences of failure are different things—each has a separate logic. Failure is violent, hectic, and crowded with people. It's also quick and sometimes sudden. The light is bad and deadly, but at least it's light. The consequences of failure are more insidious. They develop slowly but inevitably. The most aggressive people disappear and leave you with your only faithful companion—solitude. You don't even notice it at first. The telephone doesn't stop ringing from one day to the next. But the calls start to come a bit less frequently. The light seems to disappear, along with the indecency and the aggressiveness. You go from media stardom to obscurity. You start to get used to this and even start seeing benefits in it.

The qualities you need to deal with failure and the consequences of failure are different—or in some ways even completely opposite. In failure, you have to know how to handle the brutality, violence, and hubbub of the initial blow. But in dealing with the consequences of failure, you have to learn to think long-term, to maintain the hope of coming back, and to keep your self-confidence up. You can't inspire confidence in others if you don't have faith in yourself. When failure comes, everyone is paying attention to the "loser." But later, you have to know how to overcome the lack of attention. The hardest thing to do is doubtless to see it as an opportunity and a respite—a chance to draw lessons for the future. Besides, once failure is inevitable, you might as well make the best of it, and to do that you have to fight the natural human tendency to take it as an injustice. Because failure never happens by chance. There is always some cause, some essential reason, some explanation that inevitably places at least part of the responsibility on the "loser."

After stepping down as RPR leader, I was able to take some time to put these lessons into effect. I remained convinced that the French people were more ready to accept change than was commonly believed and also that the right could win elections and make such change possible. I used the following two years, during which France was still being governed by the left-wing coalition of Prime Minister Lionel Jospin, to develop my thinking on policies and politics. I remained confident that I could make a comeback, along with the right, and that we could successfully implement our ideas. We would get the chance to do so in the spring of 2002.

CHAPTER 3

Becoming a Minister

APRIL 21, 2002: THE PERFECT STORM

It would be three more long years before we on the right would get
the chance to come back to power. We succeeded, but the circumstances in which we took over—following the April 21, 2002, presidential elections—were difficult. Jacques Chirac won 82 percent of
the votes in the second round against far-right candidate Jean-Marie
Le Pen. But Chirac's election could not hide the poor performances
of all the candidates from nonextremist parties in the first round.
Nor did it hide Chirac's lack of a specific political program, with the
notable exception of public safety issues. I always thought that what
happened on April 21 was the result of the large numbers of votes
won by extremist parties as well as the poor showing of Lionel Jospin
and—let's be honest—of Jacques Chirac.

In reality, it was easy to see this catastrophe coming. For years,
the gap between the French and their political leaders had been
growing, and the signs of a forthcoming political tidal wave were
becoming apparent.

The first telltale sign was political instability. Since 1981, the
French have never reelected an existing government; this is not the

case for any of our European partners. In Germany, the Helmut Kohl government was in power for sixteen years between 1982 and 1998. In Spain, you had nine years of the José Maria Aznar government following fourteen years under Felipe González. Over the past twenty-seven years, Britain has had three prime ministers, France twelve. This is a disturbing number that belies all the speeches about the stability of our institutions. During this period, our European partners have had time to undertake reforms and make necessary adaptations.

The phenomenon of voter nonregistration, estimated at around 6 to 10 percent, is also growing, even though young people are automatically registered in the constituencies where they live when they reach the age of eighteen. It therefore cannot be attributed to their negligence or forgetfulness. Between 1998 and 2002, Paris lost 1 percent of its residents but 13 percent of its registered voters. Brest lost 1 percent of residents but 11 percent of registered voters. The number of blank or spoiled ballots rose from 0.9 percent of registered voters in 1974 to 3.4 percent in the first round of the 2002 presidential elections—one million people.

Finally, only in France has the protest vote become so significant—19 percent for the extreme right on April 21, 2002, and 30 percent including the extreme left.

If you add together abstentions, blank or spoiled ballots, and the protest vote, 56 percent of voters don't feel at home in our democracy as it currently operates. That's more than one voter out of two, whereas in 1981 the level was only 30 percent. How can political leaders expect a majority of the French to support our reforms when a majority of the French didn't vote either for us or for our opposition in Parliament?

The reason Lionel Jospin failed in the first round of the presidential elections was that he was unable to attract more than 16 percent of voters to himself and his political project, and not because leftist votes were divided among a range of rather fanciful candidates. And how bizarre that Jospin launched his presidential campaign by asserting that his project was not Socialist! His voters showed they got the message loud and clear by not voting for him. If that was the goal, it was a great success. You don't respond to voters' search for meaning by refusing to take responsibility for your political identity. The right has often been guilty of this by constantly apologizing for not being the left. It's only by building on your political identity that you can widen your base of support. Whether you're on the right or the left you should try to have the widest possible appeal.

I have worked hard to get the right to overcome its hang-ups. For a long time it seemed paralyzed by a left that liked nothing better than to posture as holier than thou. And this is why the right ended up losing much of its identity. The right found itself forbidden to talk about immigration, scolded for mentioning law and order issues, and criticized when it took an interest in education or culture, issues that were supposed to be reserved for the left. Instead of defining itself by what it was, it ended up defining itself by what it wasn't; neither right, nor left, nor center. The end result was an amalgamation of the worst of all possible worlds: too far right for the left, but not enough for the right. The right became too flexible on traditional values, and too inflexible on modern ideas. This suicidal strategy partly explains why the far-right National Front has remained so strong.

Still, whatever mistakes Lionel Jospin made, it was a disaster for Chirac to be left in the second round of the presidential election with Jean-Marie Le Pen. This was simultaneously so contrary to France's identity that no one could believe it, and so predictable that no one saw it coming.

DEALING WITH REALITY AT THE INTERIOR MINISTRY

After the president of the Republic chose RPR Member of Parliament Jean-Pierre Raffarin as prime minister, I accepted the responsibility of becoming minister of the interior. Truth be told, I was not really disappointed not to have been named prime minister, as I already doubted the president would offer me the job. Jacques Chirac fought the campaign to win and he won. He wanted to govern. To name me would have meant sharing power, and that's not his style. On the other hand, I wanted to take part in what the new government was going to do.

Why the Interior Ministry instead of the Finance Ministry, which was offered to me and which I was to end up taking two years later? First because the French had great expectations in this area and my mandate would be clear. Since my failure in the 1999 European Parliament elections, I had thought a lot about the state of our country, about our political family, about the way we governed, and about our political style. The French wanted action and this ministry needed major reform, so the basis seemed to be there to implement what I thought was urgent: the rejection of inevitability and the imposition of a results-oriented culture.

Moreover, the Interior Ministry is real life—its tragedies and passions, which come knocking constantly at your door, day and night: hostage taking, terrorist threats, forest fires, protests, rave parties, avian flu, floods, disappearances . . . it's a heavy responsibility. Not a week goes by when you don't have to make and take responsibility for difficult decisions. You give instructions that involve risks for those who are trying on a daily basis to keep the French people safe: like calling in helicopters during the November 2005 riots; allowing suspected assassin Yvan Colonna to take a walk for several hours the day of his arrest in order to finalize our plans to capture him; choosing the right moment to arrest suspected terrorists; authorizing firefighting tanker planes to resume flying after a tragic crash; giving the order to the police and gendarmerie counterterrorism units to enter a hostage taker's house. The examples are as diverse as they are numerous. You need intuition, a sympathetic ear, experience, and luck, as I discovered throughout my three years in office. But I knew already in May 2002 that the men and women who work in this ministry have a strong sense of public service and a strong concept of the public interest. These are jobs in which people risk their lives. You don't see work like this all that often: it makes the people who do it strong, appealing, demanding, and also sensitive, since they are constantly faced with human suffering. I would ask much of them. I knew they'd be there when needed and I wanted to live this adventure. They didn't disappoint me.

What is less well known is that the Interior Ministry is also the ministry of individual liberties. People tend to focus on its role in providing law and order. But they forget that this is the ministry of freedom of movement, speech, assembly, religion, and local government. This is in no way a contradiction. Security is the essential condition for freedom. But this added to my area of responsibility important and sensitive issues such as immigration, religion, and local government, especially concerning Corsica.

My first step was naturally to choose my team, especially my chief of staff. In governing, there are really only two difficult things to do—but they're very difficult. The first is choosing the right colleagues. No doubt the hardest thing to do is to know how to decide among those close to you. You have to put the right person in the right place, and if you make a mistake, you must fix it right away. This requires not only a good sense of people, but also a lot of experience in knowing what skills are needed by senior officials and finding the rare gems. I had known for years that Claude Guéant had all the qualities of a top expert. What I discovered with time and experience was that he was also a great person, and this made him an indispensable friend.

The French civil service suffers from the same lack of respect as French politicians. Like any profession, it has good and bad elements, but it is deeply honest and includes top-quality men and women. I must stress what a great advantage it is to be surrounded by women and men who have such long experience in government service. Throughout the twenty-five nights of riots during November 2005, Claude Guéant and Michel Gaudin, the director general of the National Police, were at my side. We pretty much stayed up all night for about fourteen straight nights. Guéant and Gaudin were calm, focused, determined, and professional. Of course it was up to me to make the decisions and then, even more important, to take responsibility for them. But I would never have been able to do it with such relative calm if I hadn't been constantly enlightened by their wise advice.

The second difficulty is in knowing exactly how much information you want to flow your way, given that the role of a political leader is to turn information into decisions. This is a critical issue. Too much information and you're swamped. Not enough and the decision is unsound, because the analysis was cut short. You have to delegate enough, but not too much. You must be fully informed, while avoiding being drowned.

In reality, you learn with experience to overcome these two big challenges. You can't improvise in politics. That doesn't mean that only insiders can do it, but you've got to force yourself to do it professionally, taking the time to understand in order ultimately to learn.

GOOD DECISIONS COME FROM EXPERIENCE IN THE FIELD

The very evening I took office, I went down to see the police and gendarmerie stations in and around Paris. Since then, I've never stopped going out in the field, seeking out contacts with the French people, making sure that our decisions are accepted, understood, and implemented by local agencies. I watch a lot of television. I consult numerous experts, even those who are not on my side. I go out of my way to attend the funerals of all the police officers, gendarmes, and firefighters who have died on the job. And I see it as my duty to meet with victims and their families. I do this mainly to help them. It's only if you've lost a loved one or endured great suffering in your life that you know how important these little gestures are for those who receive them, even if they might seem insignificant to those who make them.

These meetings are also special opportunities to understand

the ways in which our system isn't working and to find ways to fix it. I acted forcefully in the area of sexual crimes, missing children, spousal abuse, and anti-Semitic acts, because I have met dozens of victims and dozens of families.

All these trips in the field, inside Paris as well as out, take a lot of time, but they're indispensable. No file, however carefully prepared, can replace in-the-field experience. No official memo gives a full picture of the situation experienced by real people, as I would come to realize later on the issue of "double punishment"—deporting immigrant criminals who had already served time in prison—and as I realized right away in the so-called Sangatte affair.

Between 1999—when the French government opened this refugee center in northern France—and 2002, no minister of the left ever went to Sangatte. Not the interior minister, the social affairs minister, or the foreign minister. I decided to go there less than a month after my nomination. The situation was grotesque. Instead of protecting its borders and those of the Schengen area against illegal immigration, France was protecting British borders by holding on its territory the migrants that the British did not want to take in. These were people we couldn't send back home because they lacked identification or repatriation papers, or because there was not sufficient peace in their home countries for us to send them there without risk. Between 1999 and 2002, Britain had proposed several times that France modify its laws to make Sangatte less attractive. But the negotiations failed because France did not want to close the center. The French authorities therefore opted for the solution of doing nothing.

There was a huge contrast between this immense warehouse, sitting in the middle of fields—well maintained thanks to the Red Cross but swarming with people coming and going—and the village of Sangatte, calm and modest, whose main street was continually invaded by illegal immigrants. With the warmth and hospitality that characterize northern France, the inhabitants of Sangatte never reacted to the refugee center with anything other than dignity and responsibility. But I think they endured it with a heavy heart, and the fact that no minister deigned to come see them in three years had convinced them that the Republic had abandoned them. Initially conceived with enough space for two hundred people, the Sangatte center soon found itself welcoming one thousand to three thousand every day, and it had become a staging and rallying point known worldwide by networks of human trafficking. One of my advisers told me of having overheard someone speaking into a cell phone while at a café in Marseilles saying "Sangatte! Sangatte! Now, you've got to take them to Sangatte!" The rest of the conversation made clear just what line of business this person was in.

I will never forget my first visit to the warehouse. Three thousand pairs of eyes focused on me, imploring and threatening at the same time. They were almost all men, and none of them spoke a word of French. They were all waiting and I had so little to offer. They were quiet and yet their silence was violent. It was that day I decided to let them all go. The solution being proposed by the bureaucracy was obviously impracticable, because it was unjust. They wanted to let those claiming to have relatives in England go there and to send the others back home if possible. I could not see us making such a choice. Given that all of them had suffered—and paid unscrupulous traffickers dearly—to get there, according to what procedure, based on what proof, and on what moral basis could we do that? We took them, so simple humanity made it imperative to keep them. What was critical was to turn off the suction pump. I decided to get the British to take half of them while France would take the other half. Eventually, after an intensive diplomatic marathon between France (to get it to close the center), Britain (to get it to change its laws), and Afghanistan (to get it to accept the Afghans who wanted to return home), the Sangatte center closed its doors on December 14, 2002, fifteen days ahead of schedule.

Getting out in the field also makes a real difference in terms of making administrative action effective. A lot of laws serve no purpose, and a lot of declarations of intent get no follow-up because political leaders do not ensure their effective implementation in the field and because local agents are not always well trained and motivated. How can you imagine an immigration policy if you've never seen a prefect's office, a waiting area, or an administrative detention center? How can you develop a plan for reorganizing police and gendarmerie zones geographically if you've never gone to the trouble to understand the way police and gendarmerie units work in rural areas?

Most of the ideas that we implemented during these four years I got by going out in the field, discussing issues with our agencies, and meeting the French people. It was by talking with the units responsible for stopping the sex trade that I came up with the idea of creating a crime of solicitation. This was not in order to go after prostitutes, who are unfortunate victims, but in order to enable the police to take them in and hold them in custody while they try to convince them to identify their pimps in exchange for a residency permit. Since 2002, we have dismantled 158 trafficking networks and indicted more than 3,700 people.

It was through chatting about various things with female police officers that I got the idea of putting psychologists in police stations to calm potentially violent situations, particularly among members of the same family. The same female officers brought my attention to the pressing need to make it possible to send away a violent spouse—even before a judicial ruling—rather than having the victim feel it necessary to flee with her children, leaving everything behind, which is a situation we've in fact tolerated for a number of years.

It was by meeting single mothers and young women living in projects with no bedrooms, no computers, no worktables on which to study that some of us came up with the idea of creating boarding schools of excellence. These are inner-city residences that make it possible for the students who want to improve their lot to study in peace and quiet. This was in 2003. At the time, everyone was against this and I was mocked for having proposed to reestablish reform schools! Today, this policy is under way in practically every neighborhood.

One of the reasons for our tendency to stand still is that we wait for perfect solutions before starting to act. This is useless, because as a result nothing gets done. Personally I see only advantages in trying, experimenting, taking the local pulse, reversing course if things aren't working and moving forward if they're working well. I do not understand the controversy over the fact that I put forward two successive immigration laws. The first one had some effect. What is wrong with proposing a second one, to improve the first on certain specific points and to take it further? The 2003 law on immigration enabled us to plug the huge gaps that the 1998 law created. It gave us back the necessary tools to fight against clandestine human trafficking networks. After this goal was achieved, the second law enabled us to move from imposed immigration to chosen immigration, that is, immigration we want based on a good balance between economic immigration—which is useful for both the sending and the receiving country—and immigration for family reunification.

THE NEED TO CONVINCE

I also believe we need to explain what we do. Contrary to what is often said in Paris, the French are conscious of the need for reform and ready to work toward the common interest. But to do that you need to put forward good arguments and make the effort to convince people. Don't think it was easy for me to go to Marcillé-Raoul in Brittany after the failed TransMusicales rave in Rennes in 2002, when some 25,000 partygoers damaged public property and left the land littered with debris. Nor was it easy to go to Marigny, in the Marne, after the May 1, 2003, rave or to Larzac after the one on August 15, 2003. Each time, I had to go and explain what happened, justify my decisions on issues such as police presence and land-use permits, and convince people that it's better to give some structure to these parties than to allow them to get out of control with major risks for the locals and the participants. Each time, I was received with a mix of skepticism and ill will; each time I left with the feeling of having convinced people somewhat and of having significantly attenuated the misunderstanding and limited the resentment. This is also about the respect that citizens deserve from their political leaders.

Going out to meet the French people is demanding. But I must confess that, for someone whose life has involved carrying heavy responsibilities and accepting major constraints, it's also the main interest and true pleasure of the job. Political life constantly offers the opportunity to meet exceptional people, fully devoted to their jobs, their passions, and their lives. I can't name any of them because if I did I'd have to name all of them. I love learning about the way people live. I love learning about new fields of work, new activities. When I visit a company, I spend more time speaking with the workers than looking at the machines; this habit often disappoints the bosses and the engineers who are so proud of their technical prowess. The meetings I've had with big distributors while negotiating lower prices for consumer goods have been fascinating. I like understanding how prices are negotiated. I like understanding how department stores are organized to manage so much stock and so many goods coming and going. I like understanding evolving consumer trends. It's a tough industry. But it's a sector that creates a lot of jobs for people living in less well off neighborhoods where it's still possible to have great success through merit and hard work.

Similarly, in a completely different area, the meetings I had during the creation of the French Council for the Muslim Religion made a big impression on me. I didn't know much about Islam or about French Muslims for that matter before I took on this task. During that process, I met men and women of diverse origins and faiths. I met many tolerant and peaceful people, some of whom were professionally brilliant, but most of whom were of modest standing, as they had only recently immigrated. They were of foreign origin but profoundly French in their thinking. Actually, I feel much closer to someone such as Ali Berka—founding president of the Ali Berka Mosque, former worker at Renault (where he worked all his life), and Moroccan national who has lived in France for a number of years—than I do to a number of Parisian lawyers.

A RESULTS-ORIENTED CULTURE

This method, supported of course by major legislative and structural reforms, produced results. In four years, crime fell by nearly 9 percent, compared with a rise of 14.5 percent from 1998 to 2002. This represents 300,000 crimes avoided per year, and more than a million victims spared. The rate of resolved crimes rose by 9 percent compared with a drop of 3 percent during the previous period. The number of people charged grew by nearly 30 percent. The number of illegal immigrants expelled from France doubled. Finally, in four years, we went from eight thousand deaths on French roadways to fewer than five thousand, and we prevented thousands of people from becoming disabled in auto accidents.

Beyond the numbers, we endowed our security services with the tools necessary for long-term effectiveness. The database of digital fingerprints rose from 400,000 entries to 2.3 million, while the number of DNA samples rose from 1,000 to 220,000. This information is useful for identifying the guilty, but also for exonerating the innocent. In a modern democracy, technical proof is much better than confessions and assumptions. Similarly, we have made the conditions of police custody more humane and installed cameras in every police vehicle. It is in everyone's interest for the police to work in calm and secure conditions, protected from the risk of things getting out of control and useless accusations being made. We created regional and multidisciplinary intervention groups—with expertise in police work, fiscal matters, and customs—to dismantle networks and gangs, especially by going after their assets and visible signs of wealth. We changed our approach to inner-city violence. Maintaining the status quo just wasn't good enough for me. From now on, we bring people in for questioning. Four thousand seven hundred people were thus arrested during the November 2005 crisis. This change in our method explains how we got through these trying events without any deaths, either on the side of the security forces or on the side of the rioters.

We also took the measures necessary to be better protected from terrorism and to get rid of the problem of hooligans. We significantly reformed immigration policy. France is now an open country, but it's free to choose whom it wants to welcome into its territory. Finally, we are developing a crime-prevention policy. For the first time, our country will have a comprehensive policy of prevention.

The purpose of this chapter is not to assess my work as interior minister or finance minister. There are other places to do that. My goal is to show, with examples, what a minister can accomplish in a few months with will, determination, and imagination. Nothing we did was easy. Many, including some within my own political family, tried to stop us. Nothing was done by snapping our fingers. We put in endless hours of work over many months. But in the end, what we accomplished was far-reaching and valuable. Though I must admit that the left had given us a lot to work with by doing nothing for five years!

Many say that government bureaucracy is ineffective and out of touch with the realities on the ground. They say it's tied up in red tape and that top civil servants do not do what the ministers want. The red tape charge certainly holds when you consider that a civil servant has to take a competitive test in order to move from the General Tax Office—which is responsible for calculating and collecting certain taxes—to the Public Accounting Office, which is responsible for collecting other taxes. But what bureaucracies need most is leadership. You have to set their goals. You must evaluate their performance. And you've got to take that performance into account when you allocate resources and make personnel decisions.

At both the Interior and the Finance Ministries, I always saw the top civil servants as my colleagues. I think the size of ministerial staffs should be reduced so that top civil servants have direct contact with ministers. I also think—for the same reason—that ministers should be allowed to choose their top associates. Only in France do you have this extravagance of the president of the Republic nominating seventy thousand officials. As for private staffs, rather than systematically duplicating directors with younger colleagues who are eager to take their jobs, it makes sense to recruit the kinds of people you don't normally find in the bureaucracy—such as researchers, intellectuals, and people from the private sector. I've always thought that what matters in a private staff is not the number of colleagues but their quality and their courage.

I have a preference for people who get involved rather than those who like to watch. That's why I never saw anything wrong with civil servants wanting to get into politics. In fact I always sought to promote prefects who were part of a ministerial staff, even if it was as part of a government that I did not support. In my mind, they had the merit of having taken risks and acquired experience that an exclusively administrative career doesn't offer. That led me to name so-called leftist prefects to senior positions. I never came to regret it. When you go back out into the field or back to the central government, it is a clear benefit to have had experience on a ministerial staff.

We also need to change the way we staff the top levels of our civil service. For thirty years, senior officials have been leaving to take high-level jobs in the private sector. We must reverse this trend so that future ministers can include top-quality colleagues who have public and private sector experience. Why not propose one hundred contracts beyond the normal remuneration categories so that the most effective private sector leaders come back to public service? We would have a lot to learn by tapping into this experience and renewing the senior civil service.

One method I often use to keep my teams focused on results is to set deadlines. As soon as one meeting ends, I often set the date for the next meeting. This provides healthy pressure to seek constantly to be effective. If I hadn't announced in summer 2002 that Sangatte would be closed before December 31 that year, it would still be open today. If I hadn't given distributors and producers fifteen days to reach an agreement on lowering prices, there would never have been an agreement. When you really want to reach an agreement you don't need three months. After the intercommunal strife in Perpignan, or the case in La Courneuve where an eleven-year-old boy was killed by a stray bullet, and in many other circumstances as well, I was always determined to review all the announcements and promises I had made. And I always came back with results. In La Courneuve, a year after the killing, the rate of solved crimes rose by 44 percent, three Grandes Écoles (Polytechnique, l'Essec, and Supmeca) developed tutoring and school support programs for local students, two boarding schools were being set up, and 196 youths had found jobs. Similarly, if I hadn't gone out of my way to highlight in press conferences what we were doing, we wouldn't have had the same effect because we wouldn't have provided the required stimulus. Finally, for months, I met regularly with Madame Erignac, the widow of the prefect assassinated in Corsica in 1998, to review where we were in the search for Yvan Colonna, who was suspected of killing her husband. Finding Colonna was a priority for me, one to which I and our police and gendarmerie forces devoted constant energy and unceasing determination. Our approach left nothing to chance. His arrest was certainly one of the most intense moments of my tenure as interior minister. I took his being free as a slap in the face of the Republic. I was in Carpentras when I was finally able to tell Madame Erignac the news. This admirable woman who had suffered so much was then able to begin her mourning. It was about time.

EFFECTIVE DIPLOMACY

At the end of August 2002, I decided to apply the results-oriented approach to our diplomacy. Doing so ruffled more feathers within the felt-covered halls of the French Foreign Ministry than within the government of Romania.

At the time we were having great difficulty with the Romanian government over illegal immigration and human trafficking. I saw no alternative to going to Bucharest to work with the Romanians toward a common approach. The draft agreement that the Quai d'Orsay prepared for me was useless. No commitments from the Romanian side, no commitments from the French side. There were grand declarations of principle, words, mutual congratulations, and pledges of friendship—but nothing concrete.

When I arrived, I proposed to my Romanian counterpart, and particularly to the brilliant prime minister at the time, Adrian Nastase, that we set aside the meaningless and empty words and that we draft a new declaration. This was not normal diplomatic procedure, but I wound up getting the following from Romania: police cooperation to dismantle the illegal immigration and human trafficking networks; modification of Romanian law to punish in Romania traffickers using prostitution networks in France, particularly by confiscating their assets; assistance in identifying and repatriating Romanians living in France illegally; and the creation of a program to welcome and reintegrate minors who might be apprehended in France and sent back to their families in Romania. For its part, Romania got from France the right to come and look for Romanians who were living illegally in France, and French cooperation in helping to create border police. All that was left from the initial declaration prepared by the Foreign Ministry were the signatures and the date.

Diplomacy is a difficult art. But saying nothing, asking for nothing, and offering nothing can lead one to despair. Two future partners in the European Union had the right to expect concrete and precise commitments from each other.

TAKING TIME TO REFLECT

During the months following my arrival at the Interior Ministry, I was criticized for going too fast, going too far, and even being too much. These are critiques that have been leveled at me at every stage of my career, ever since I was very young. I was too aggressive, too ambitious, too hungry. And it's true, I love life. I love it so much that I've always wanted to live it in full, seizing every moment. I've always been surprised by those who advise me to "take my time." As if time were ours to take and you could modulate it as you like. For them, it was always too soon for me, even if it would soon be too late! I have seen so many people who, by waiting, never did anything at all, that I've been inspired by their example to do the exact opposite. I would rather take risks by daring to do things than regret that I was not able to seize the opportunity when it was there.

Similarly, every time I have been responsible for something, whether at the national, regional, or local level, I always got to work from the first minute of the first hour of the first day, and I did not let up until the last. You don't get to be a minister all your life, and here, too, I've known too many people who, when it was time to go, were still thinking about what to do!

I don't feel as if I go too far; indeed I think most people think that their leaders don't go far enough. I like to build, act, and solve problems. My weakness is thinking that there's always a solution, a possibility, and room to maneuver. I believe in will and determination. I don't resign myself to failure. I love tenacity. I rarely if ever give up. I believe that everything is earned, and that effort always pays off. These are my values. This is what I'm made of.

I work hard, because contrary to the idea people have of me, I have lots of doubts. People see me as having strong convictions, and in a sense they're right. They are the result of thirty years of political life, which is not insignificant. Yes, in thirty years of public life, I have become convinced that to conduct a balanced policy it is necessary first to win over one's natural base—to reassure it and encourage it to open up. Yes, I'm convinced that deep down in French society there is a strong demand for the restoration of certain values of the republican right: work, respect for authority, family, and individual responsibility. And I'm convinced that the reason the right has been losing for years is that it regrets not being the left. Yes, I'm convinced that no country in the world can get by without effort, and that France—notwithstanding its undeniable merits and prestigious past—will become a thing of the past if it doesn't take the steps necessary to adapt to the changes taking place in the world.

But the strength of my convictions is not in any way in contradiction with the doubts I always have. I try never to take any alleged certainties for granted. I need a lot of time to understand and analyze a situation. I write my speeches by hand, and often with as much difficulty as meticulousness. I meet with a wide range of diverse interlocutors to make sure I know all the possible angles of a difficult issue. I'm not at all impressed with set habits, rules, or reasoning. I would rather think twice before deciding than continue to hesitate after making a decision. Doubt is a part of judgment; hesitation doesn't get you anywhere.

I therefore reflected for a long time before I made the decision to leave the government to become leader of the Union for a Presidential Majority (UMP) in 2004, when I was minister of finance. The president of the Republic had insisted that I either leave the government or give up on the presidency of the party. This was a special rule, created just for me and this particular circumstance. It was really wrenching for me to leave the concrete activities of the Finance Ministry to take over leadership of my political family. I thought about it for several weeks. My team was itself divided on the matter. But ultimately the decision became obvious. I had to take over the organization that would be decisive for the next presidential election.

I also thought carefully before coming back to the government one year later. It has been said that I came back to the Interior Ministry to protect myself from the scandals that people were stirring up against me. There were, alas, some scandals. I'll discuss them later. But they were not the reason I came back. I tried to understand what the French people were expecting from me. They didn't expect me to start campaigning two years before the election; they wanted me to get back to work on public security without delay. Actually, all my friends were against it, fearing that I would be dragged into the turbulence that's an inevitable part of government. I made clear to them that it was my life, that I was the one who would pay the price in terms of hours of work, stress, and pressure. I made clear that I had thought more about this than anyone, and this was how I felt. Édouard Balladur, Pierre Méhaignerie, and even former president Valéry Giscard d'Estaing—all people I like, feel close to, and respect—thought until the last minute that they could get me to change my mind. But I was sure that I had devoted enough time to thinking about it that my doubts had disappeared. I was sure that I had to come back, rather than remain on the outside where all I could do was talk. Seeing the limits of François Bayrou's strategy—staying out of government and running for president— ultimately convinced me that it was not for me.

Finally, I need time to let a decision mature. People say I'm instinctive, but I am organized. I can be wrong. I have been wrong. But the decisions I make are rarely made by chance. I have always prepared for those decisions with careful reflection. I try to put together a strategy and to stick with it. The period of doubt is the time to get organized. I'm neither apologizing for it nor bragging about it. That's just how it is.

TOTAL COMMITMENT

It is an opportunity, an honor, and even a privilege to have the confidence of voters, to be able to change people's lives as a minister can, to live a really interesting life, and to serve your country— especially one like France. I never understood why some people complain about being in politics.

After all, nobody is forced to do it. There are enough people who want to do it that those who find political life too difficult can just give their spots to those who so desperately want to take them. And beyond the challenges and the disappointments, I am fully conscious that I am living, day after day, the dream I've had since I was twenty: to build. When you're young you dream about what your life will be like. As an adult you try to turn those dreams into reality. Matching yesterday's dreams with today's life should make it possible to be happy.

This personal accomplishment does not, however, come without effort, sacrifice, and difficulties. This is the fate of everyone who seeks to take his or her passions to the limit. Nothing precious comes without giving it your all, and this commitment has a price. That's what gives legitimacy to your ambition. That's what makes it noble.

Politics led me to make painful choices, even if I try to reconcile this aspiration as well as I can with the needs and expectations of those who are most dear to me. To be at the top level demands total, permanent, and all-consuming commitment. The price is high for your family and for you. From the outside you can't really imagine what it's like. Over the years, I've come to understand just how heavy a burden it could become.

C.
___

I thought hard before I wrote these lines. Wasn't I just going to reignite the controversy? Wouldn't I open myself up to the criticism of having left my family exposed? Wasn't my family itself going to suffer again? All of these questions turned in my head for a long time. In the end, I've decided to talk because others have already done so and will continue to do so. So at least now the person most directly concerned can give his views. After all, this seems only appropriate. I want to put an end to a debate I suffered through, that hurt me, and from which I think I learned something. Finally, since I recommend authenticity in public debate, it seemed to me more honest to begin by applying that principle to myself.

Millions of people have gone through what my family went through. Their suffering, doubts, and hopes are the same as ours. These are eternal stories about men and women. They're stories about the difficulties of sharing a life and about the power of love. The only thing that's different for my family is the pressure of public life, fame, and publicity. Everything between men and women is complicated, but when everything is public every little thing that happens in daily life becomes a huge deal. To handle this, you need a degree of energy that I didn't know I had. It's the biggest price you pay in modern political life. I say "modern" not only in the sense of "modernity" but also meaning "these days." This evolution toward transparency in public life, unimaginable only ten years ago, has become inevitable today. So you might as well deal with it head-on and not try to dodge it.

C. I write "C." because still today, nearly twenty years after we first met, it moves me to pronounce her name. C. is Cécilia. Cécilia is my wife. She is part of me. Whatever challenges we have faced as a couple, not a day has gone by that we didn't talk. Really! We didn't want to betray anyone, but we're incapable of being apart. It's not that we haven't tried, but it's impossible. We finally realized that it was vital for us to speak to each other, listen to each other, hear each other, and see each other.

I've often been criticized for wanting to make a show of my relationship. I understand this criticism and in no way want to shirk my responsibility. But I would like it to be understood that nothing was staged. Everything was sincere and true. We were made for each other. Showing my public life was also showing my private life because they were the same thing. Nothing, and I mean nothing, was made up.

When I realized that I had exposed Cécilia, the damage was done: too much pressure, too many attacks, not enough attention from me. At the time, our relationship didn't hold up. Then all hell broke loose. Everything was affected. Even today it's hard for me to talk about it. I never could have imagined going through something like this. I never could have imagined being so devastated. In this hurricane, I found two consolations. One was Cécilia, who suffered just as much, but who always believed in our future together, strange as that may sound. The other was the many people we didn't know who shared with us their own, similar experiences on the brink. Sometimes it's said that politicians don't know "real life." Well I can tell you that I've experienced a bit of it now! Maybe this even forced me to show a bit of my human side that was no doubt lacking. This kind of ordeal is in no way a matter of pride or jealousy. It's more deeply and simply about love. The challenge is the emptiness you feel, not the wounded pride.

Today Cécilia and I have gotten back together for real, and surely forever. I'm talking about it because Cécilia asked me to speak for both of us. She wanted me to be her spokesman. She could no doubt have said it better than I, but by asking me to do it she showed her modesty, her fragility, and maybe also her confidence in her husband.

We won't talk about this anymore because, even though I'm only saying a little bit, I hope the reader will understand that this is a lot, given how important C. is to me.

In particular, I hope that however famous we may be, everyone out there will understand and accept that our story is simply that of a man, a woman, and a family. We do not deny our mistakes but ask to be respected so that we can calmly continue along the path of a life that we now know is not easy for anyone. A life in which we, like anyone else, need love. I now know this to be so precious that it must be protected. The past will serve as an eternal lesson.

CHAPTER 4

Doing Things Differently

POWER STILL EXISTS

I'm not someone who thinks that power no longer exists, that it's lost its value, or that it's somewhere else. Like somewhere other than politics. Put differently, in my view nothing is inevitable. This is what I have tried to show by committing fully to my work at the Interior and then the Finance Ministry. There are always angles of approach, even if they're very narrow. It is more than ever possible and necessary to act.

To be sure, the context is evolving. Power is not exercised the same way as in the past. Communications have become primordial, less because of the development of the media than because of the way that society has evolved. Citizens, the French as well as those of other democracies, are better educated and thus more demanding. In between elections, they don't blindly put their trust in the governing majority. They want to know and understand the intentions of the government on a daily basis. It is impossible in our era to act without informing, explaining, communicating, and seeking to convince.

To go even farther, thirty years ago, you acted and then you explained. Today it's the opposite: you get authorization from "public opinion" to act only when you've explained something well. Communication has become the prerequisite for action. It's the first stage. You can no longer distinguish between form and substance. They make an inseparable whole.

If you don't go through this stage, it becomes impossible to act. Shortly after my arrival at the Interior Ministry in 2002, when I was putting forward my draft law on internal security, *Le Monde* ran a headline saying "Sarkozy Goes to War Against the Poor." I guess they also decided to get into the communications business. I fought every step of the way. I didn't want to apologize or to let them dominate me with this attempt to prevent me from acting. I found it insulting for the millions of people who have great social difficulties but who are not therefore criminals. It was by winning this first battle with the media that I was able to implement a new security policy for the French people.

Similarly, it is true that European integration—and our international commitments more broadly—limit our national room for maneuver. Certain reforms are impossible if we don't have the agreement of our international partners or EU institutions. Economic globalization means that national decisions must always follow the dictates of the mobility of factors of production, and especially capital. The increasingly open circulation of ideas, information, people, and goods creates new problems such as migratory pressure, environmental degradation, and terrorism.

Still, political leaders are capable of action and should stop using these things as an excuse for inaction. Globalization, Europe, and our international commitments do not replace political action. They just add new constraints. It's up to us to figure out how to deal with them. Do people think it was easier during World War I, the Depression, the defeat of 1940, reconstruction, or decolonization?

Political leaders must adapt to the new constraints and find room to maneuver wherever it exists. The conditions of exercising power are changing, but not its purpose, which is to act. The primary mission of a politician remains to give renewed hope by showing that it's possible to influence events. In this way, what seems impossible becomes possible, and what seems inevitable can be questioned. It's all too easy to plead that you can no longer control events, even while you're running for office. But why, then, seek to exercise power if it no longer exists?

I also want to note that many current problems in France have nothing to do with European integration or globalization. This is true for the rising insecurity and crime from 1997 to 2002. It's true for the case of the suburbs that we allowed to sink into despair and lawlessness. It's true for education. After the undeniable success of extending and democratizing education in the 1970s, the school system today is in crisis. Many students are failing and inequality is rising. On all these issues, neither Europe nor globalization prevents us from acting. The same is true for our criminal justice system, which is too soft on criminals, too hard on the innocent, and too distant from victims—and for our research sector, which falls farther behind every day. Ditto for our health care system, which is ever more expensive, as is inevitable, but which is not thereby able to deal with such common issues as addiction or tragedies like cancer. Patients often have to wait several weeks before receiving care and treatment for the suffering that this terrible disease inflicts on them. Is this acceptable in a country as rich as ours that spends so much on health care? I could give many more examples of these delays and malfunctions in French society that no international constraint prevents us from changing.

REDUCING PRICES: FAIR ACTION IS POSSIBLE

When I decided in April 2004 to negotiate a deal with producers and distributors to lower prices in the big-box retail sector I was put under great pressure by the business world. They told me that the stock price of Carrefour—the second-largest retailer in the world, with 45,000 workers in France—would crash, making it easy prey for Wal-Mart, the world leader known for its brutal and unacceptable business practices. The same people claimed that lower prices could threaten the viability of a number of businesses in the agricultural sector. The producers and the distributors, normally at each other's throats, agreed to oppose any change. One distributor, whose name I won't reveal even though I know it, circulated an unsigned document to politicians, journalists, and opinion leaders that claimed—using the entirely irrelevant example of the Netherlands—that a 1 percent drop in prices would result in one million job losses.

In France we call anonymous memos *notes blanches* ("white memos"). When I became interior minister in 2002, one of my first decisions was to eliminate the use of *notes blanches* by the domestic intelligence agency and insist that every memo be signed by its author. We're not living in 1940. You can speak freely in this country. Whoever has something to say should take responsibility for saying it.

The risks involved in the effort to lower prices might have stopped another minister in his tracks. But the worst thing was the stance taken by the Finance Ministry's bureaucracy. Their view was that there was no point in applying pressure. For the bureaucrats, the Finance Ministry hadn't had any real power for years—at least when it came to setting prices.

I had a different view. Of course I had no legal authority to lower prices. But I was sure that with the support of public opinion—that is, consumers—the producers and distributors couldn't say no. Ever since the creation of the euro, our measures of inflation underestimate price increases. Some tried to play the same games with prices that they did with security: distinguish between price increases and the *feeling* of price increases. The French were not fooled. So we had to act: doing so was fair to consumers and useful for the economy, because it boosted consumption and therefore helped both producers and distributors. In fact, thanks to our actions, prices of consumer goods have fallen consistently since September 2004, much more than the 2 percent foreseen in the agreement. Carrefour has not been bought by Wal-Mart. No agricultural company has gone bankrupt, and the idiotic 1996 law, which is what led to the inflation in the first place, is being modified in a way that serves consumers' interests. God knows how isolated I was at first.

The story has been told that on that famous June 2004 night when we reached an agreement, I threatened to go on television to talk about the big industrial groups' pricing practices if they didn't agree to what we wanted. The story is true, and this was exactly the thing to do. The Finance Ministry is the ministry of all French people, not of certain companies. What right did I have to let certain big industrial or distribution groups continue to get excessively rich on the backs of the French?

It's not the absence of power that undermines politics and delegitimizes those who wield it. It's the fact that they no longer wield the power they do have. Here's the heart of the problem: a sort of abandonment of political will that evolves into a failure of imagination. Since nature abhors a vacuum, the longer this situation endures the more nonpolitical forces will take advantage of it.

It has become fashionable these days to criticize the skyrocketing profits of certain companies listed on the CAC 40 (the leading French stock-market index). But rather than spending our time denouncing everybody in grand speeches of no practical importance, let's look at the issue more closely. If these profits are undeserved, as in the case of the big stores in the retail sector, let us act and undertake reforms. But if they're not undeserved, let us congratulate our companies on their success.

By lowering the prices of everyday consumer goods, I showed that the finance minister could still affect the daily life of the French. This was also a way for me to avoid the trap that the Finance Ministry would have been for me if I spent my time at grandiose international symposiums on theory instead of focusing on concrete national issues. A lot of my friends didn't think I should take on the responsibilities of the Finance Ministry. They thought I risked cutting myself off from the daily preoccupations of the French, which had been my specialty since 2002. I quickly pushed aside these fears, which in any case resulted from ulterior motives as much as from sincere and friendly thoughts. This was because for me, whatever the ministry, it's always possible to act. Whatever position of responsibility you're in, you can and must apply the same principles with the same will. The problems change but the issue remains the same: convince the French and, by taking action, restore their hope.

TAKING INSPIRATION FROM OTHERS

Our European partners live under the same constraints as we do, but they have managed to adapt without losing any part of their identity, and they're living pretty well. We should think about their example. The most impressive case in my mind is that of Great Britain, which, it will be remembered, in the late 1970s seemed to be completely left behind, with a GDP 25 percent less than that of France. It was a country that saw itself destined to endure unemployment and industrial restructuring, or even to accept the disappearance of its major industries. Do the French ever wonder why the British are buying our houses in the Dordogne, Périgord, the Luberon, Savoie, and so many other regions? It's simply because Britain's GDP is now 10 percent greater than France's and the British standard of living is higher. I have nothing against the English, who are our friends, but it is not my ambition to see France's most beautiful villages set aside exclusively for British vacationers!

What's worse, London has now become the seventh-largest French city. It has attracted, practically to the point of saturation, thousands of young French people who go to live there, including my daughter. It apparently seems easier to succeed there than in France. Or worse, it seems that success has become so shameful in France that a young person who wants to succeed must leave. I do not accept this minimalist vision of what France has become. I am not resigned to seeing the most dynamic and active part of our society—our youth— abandon France. A million French people have gone to live abroad over the past few years, a loss almost equivalent in absolute terms to the losses of World War I (1.3 million French deaths).

The case of the Scandinavian countries is also telling. In the late 1970s and even the 1980s they were examples of what not to do, with their exorbitant levels of taxation and unsustainable debt, but now we admire them: Denmark, for its "flex-security," which marries labor market flexibility and job security; Finland, for its education system (which was ranked first in the Organization for Economic Cooperation and Development [OECD]'s Program for International Student Assessment); Sweden, for its spending on research and development (second highest in the world), promotion of women, administrative model, environmental policy, and sports.

These successes cannot be reproduced exactly in France, but they should make us think. They didn't come about by chance. They're the product of internal political decisions that show that states can still act. In short, why should we not believe that we can accomplish the same things? In what way would we weaken our own identity by enriching it with others' successes?

When I think about economic globalization and the growing interdependence of states I don't see a reason not to act—rather, I see a reason to innovate. I believe we really must open up French political life to the experiences of others. Our public debate is not sufficiently enlightened by others' successes, initiatives, or failures. I have often wondered why we have this propensity to try everything that doesn't work and to be afraid of trying everything that works elsewhere. We could learn so much from the Danes, the Spanish, the British, the Germans—and even the Americans!

To refer to the policies carried out in other countries is not to impose a foreign model on ourselves. "Sarko the American" is a comment you hear in France that's meant to suggest that I want to transform the French social model into the Anglo-Saxon model. This is just the sort of jab that's so common in French politics, designed as it is to prevent thought and action—or I might even say designed to kill. And it lumps together everything that can possibly make me seem like a henchman for the inequalities and excesses of the United States: belief in free markets, my point of view on affirmative action, my institutional proposals, my book on religion—though it tends to overlook that the first word of the title was "Republic."

If I were in love with the American model I'd go and live in America. This is not the case. I admire the social mobility of American society. You can start with nothing and become a spectacular success. You can fail and get a second chance. Merit is rewarded. There are fewer social codes than there are in France. You're not judged immediately based on the way you say hello or what your name sounds like. On the other hand I'm not a great fan of the American social model. Social insurance is insufficient and unequal. I do not accept that someone must receive substandard health care or no health care at all, just because he or she is poor. I do not accept that someone can live in permanent fear of getting sick because there is no social insurance.

As for affirmative action, let's discuss it. Americans had the will to address the issue of certain disadvantaged minorities, especially the black population, with a policy called "affirmative action"—what the French call "positive discrimination." In France, some have caricatured this policy by reducing it to the issue of quotas in universities. In so doing, they forget two things. First, affirmative action initially took the form of reserving public contracts for companies whose personnel reflected racial diversity. Second, quotas were ruled unconstitutional by the U.S. Supreme Court in 1978 and then limited to certain very specific cases. Far from being a simple matter of quotas, affirmative action in the United States was a great awakening and a sign of political will. A society opened its eyes and decided to make individual equality not just theoretical but real. It wanted the diversity of American society to be represented in all sectors and at all levels of economic, political, and social life.

This policy has hardly solved every problem. But today there's a black middle class, a Hispanic middle class, and an Asian middle class. Major figures from minority populations have risen in every area of American political, economic, social, and scientific life, as well as in the American media. These include Secretary of State

Condoleezza Rice; former secretary of state Colin Powell; Supreme Court Justice Clarence Thomas; Secretary of Commerce Carlos Gutierrez; and Attorney General Alberto Gonzales. They also include Kenneth Chenault, the CEO of American Express; Richard D. Parsons, the CEO of Time Warner; Stan O'Neal, the CEO of Merrill Lynch; Zalmay Khalilzad, the U.S. ambassador to Iraq; Fareed Zakaria, one of the most visible and well-respected foreign policy experts; and many others. Like it or not, this is not yet the case in France, for any of our recently arrived minorities.

Affirmative action in America is an experience that could be an inspiration for us. Does it require quotas? Not necessarily. It's mainly a sign of political will that should wake us up. France has a huge problem integrating immigrant youth into society. It's a problem that has plagued France for years and that prevents it from dealing with other challenges. But it's a problem that can be dealt with if France gets motivated to do so.

In 1598, in the Edict of Nantes, Henri IV ended nearly forty years of barbaric religious wars and guaranteed peace for nearly one hundred years. The Edict of Nantes was not merely a bunch of words. It conferred concrete rights, amazing for the time: freedom of religion; civic equality; the opportunity for Protestants, who made up 5 percent to 10 percent of the population at the time, to hold top offices without distinction or discrimination; the right to go to school, university, and hospices. It gave the same rights to Protestants to be judged by special tribunals including Protestant magistrates, and it gave them 150 publicly funded protected places—châteaus and fortified villages that could protect them. In 1791, the French Revolution emancipated French Jews and recognized them as French citizens, giving them the opportunity to participate in the country's economic and social life. This policy was begun by Louis XVI and extended under the Empire, which made the Jews of France the strongest defenders of the French model of integration.

To each era its own answers, but let them be real, assertive, and determined answers. To take an interest in American affirmative action is in no way to renounce either our history or its model of integration. It doesn't mean we want either American-style ethnic communities or urban segregation. It just means looking for pragmatic solutions to a burning problem for French society. It's a call for an awakening. It's convincing people that there are more risks in huddling behind our beautiful principles than in trying to speed up the integration process for those whose skin is a different color from that of the majority. Affirmative action, in any case, must be only temporary. The legitimacy of any particular, additional measures by the Republic toward this or that category of people or territory is to restore balance. Once this balance has been restored, the extra effort should come to an end. What undermines the French model of integration is not letting young Moroccans and blacks become lawyers, engineers, journalists, bankers, businesspeople, or ministers; it's continuing to tolerate letting a top graduate remain unemployed after a year because he's black or Arab, when all his classmates have found jobs.

The French want to be French and can only be French. It would make no sense to copy a foreign model that would not work in any case. France must find the energy to succeed in its own traditions, not by imitating others. It must find the strength to overcome obstacles in the enduring qualities that have forged its character and its identity. Only in this way can it figure out how to enter the new millennium without renouncing what it is but also without fear of the future.

The French are attached to their values—as they should be, because these values have inspired the world. I would like my compatriots to be confident about France's future. France is not a museum. It can again become an example to others. This would certainly require some self-examination, some updating of our values, which are not always the ones we think they are—because in order to be an example you've got to be exemplary! Our social model is no longer exemplary, nor is our system of integration, and our economy even less. We confuse equality and egalitarianism; solidarity and special help; justice and leveling off; patriotism and nationalism.

France must again become the homeland for work, merit, responsibility, and fraternity. It must be the country where social advancement is possible, encouraged, and desired by all.

DEALING WITH EUROPE

For years we have explained to the French that if nothing can be done, if nothing can be changed, it's because of Europe. Voters took us at our word and voted against the draft European constitution in 2005. Europe doesn't come out of nowhere; it's made up of states. It's up to the states to act if they want the European Union to work differently.

This is what I tried to do in the areas of immigration and security. Issues such as security, asylum, and immigration are increasingly dealt with at the EU level. This is mainly a legal reality, something the member states chose. It is absolutely necessary. The opening up of internal borders has made it easier for organized crime networks to circulate freely around Europe. Whereas an international fact-finding mission—as I know all too well—takes several months to come back from Italy, terrorists can cross all of Europe in seventy-two hours. Only harmonization of our police and judicial procedures and coordination of our police forces can compensate for this ease of movement. Any foreigner who enters an EU member state in reality also enters all the other states that have eliminated their borders with other EU members. It makes no sense for each country to have its own immigration policy, with some more restrictive than others. Finally, only Europe is able to put in place a policy of development aid that can meet the needs of the countries of origin. Thus EU countries have every reason to work together.

Unfortunately, they don't. In 2002, Europe's immigration and security policy still followed—and this mostly remains the case today—the unanimity rule. This means that all EU countries must agree before any common action or changes in EU legislation in these areas can take place. This rule impedes the effective functioning of the Union. Not only is it rare for all twenty-five members to agree on everything, but even worse, the unanimity rule blocks negotiations from the very start. With such a system, no country has an interest in starting any discussion, because it knows that it ultimately can't be forced to do anything. The best way never to reach a compromise is to say that no compromise is needed.

The unanimity rule, which is supposed to protect the vital interest of each EU member state, has over the years become the source of enduring obstacles. It stalls Europe in ways that exasperate Europeans and make them feel distant from an essential cause. Only majority voting can end the delays of a decision-making process that is incompatible with the kind of quick reactions needed in the area of security policy. The countries that do not want to move forward should be allowed to keep their own laws. The countries that want to go slowly should not be allowed to block those who want to accelerate. The enlargement of the EU from fifteen to twenty-five members made this situation worse.

Whereas security and immigration were the two main issues in my ministerial portfolio, I quickly understood that acting at twenty-five would be impossible, or at least very difficult. Each member state has its own challenges. Spain, for example, has to cope with as many illegal immigrants per day as Cyprus does per year. Countries without major immigration issues, mainly in Northern Europe, support generous principles—for example, when it comes to marriage or family reunification—that countries facing bigger challenges from immigration cannot be expected to adopt.

Faced with this sort of stalemate, I could have just stood there and shrugged my shoulders. I could have gone on French television and said, "Nothing I can do, Europe prevents me from acting." Many of my predecessors didn't hesitate to take this approach. But that's not what I did. On the contrary, in November 2002, at the Franco-Spanish summit in Malaga, I proposed to the Spanish that we create a group of five big EU countries (Germany, the U.K., Italy, Spain, and France) to push forward a much more ambitious European security, asylum, and immigration policy. Thus was the idea of a "G5" launched.

It was never my intention to create a "Directorate" or to impose big countries' decisions on the smaller ones. The European Union operates according to precise legal rules, which the G5 does not have the power to change alone. Instead, I was convinced that the security and immigration policy needed a new political impetus, and that this impetus had to come from the five big countries most exposed to problems of immigration and insecurity. I also knew that the personal relationships that we could develop among ministers could help us overcome obstacles. When an official tells an EU meeting that his government agrees to negotiate a particular directive, it is easier for his state to back out or play for time than when the commitment is taken by five ministers in the framework of a relationship of confidence, or even friendship.

To be honest, I got fed up with the meetings of the Justice and Home Affairs Council—attended by justice and interior ministers—owing to the interminable discussions of technical subjects that were of interest to only a few people. They never led to any operational decisions. At least the five big countries had the same problems to overcome and the same urgency about doing so. I much preferred the pragmatism of our meetings at five to the diplomacy of our meetings at twenty-five.

By proceeding in this way, all I was really doing was taking a cue from the way the Franco-German couple functioned for years. Founded on the strong will of reconciliation and sometimes on the friendly relationships of its leaders, the Franco-German couple provided the impetus for new European policies for a long time. It signaled that negotiation was possible because a compromise had been found between the two countries that were generally on the two extreme ends of the European debate. The EU today is bigger and more diverse. The French and German points of view no longer come close to covering the full spectrum of possible positions, especially compared with Great Britain and the Central and East European countries. This is why I think that, while Franco-German agreement remains necessary, it's no longer a strong enough motor for Europe today. This seems obvious to me.

It took me some time and much determination to convince my foreign counterparts as well as the president of the Republic to accept the principle of the G5. They were all afraid of how the smaller countries would feel about being "left out." Ultimately, we were able to overcome all these obstacles, and the first G5 meeting took place on May 16, 2003, in Jerez de la Frontera, Spain. Sadly, two days before this meeting, the city of Casablanca was ripped apart by deadly attacks, the worst ever on Moroccan soil. The need to act quickly and in a concerted manner was now all the more apparent. Since this date, the G5 has formulated a number of proposals in the area of security and immigration that have been taken up in turn by the full EU at twenty-five. Similarly, we started taking some common actions, especially in the area of counterterrorism. We now share what we know about terrorist networks, and we have created at the G5 level a DNA database for counterterrorism. The September 11, 2001, attacks and the March 2004 Madrid attacks have demonstrated how essential it is for democratic countries to quickly exchange information. In both cases, in fact, good cooperation among intelligence services came close to thwarting the criminals' plans. How many attacks are thus prevented, thanks to such cooperation?

The G5 is the proof that it's still possible to act in Europe. And I think it should develop further: first by taking in Poland, which with a population of 39 million is certainly a big European country. And then it should link up with the other European states as long as they wish to move forward in the areas of immigration and security. This way no one will feel left out. Everyone who wants to act will be able to participate, and at the same time a genuine EU policy can be put in place.

Creating the G5 to break EU immigration policy out of its impasse was my own idea. Many later tried to dissuade me from starting it up, if only by expressing doubts about its feasibility. It was the refusal to accept standing still that helped me overcome the strongest reluctance and the most discouraging attitudes. Nothing is stopping us from applying the same imagination and determination to other areas of EU action.

The French criticize the EU for being abstract, for not developing concrete policies that European countries need—for example, in the areas of fighting against job loss, development assistance, and research and innovation. This is true, but they forget—or rather no one reminds them enough—that the EU is more than anything a grouping of states. The EU is stuck because its member states are stuck, particularly the Franco-German couple, and especially France, which continues to lose influence. One statistic suffices to illustrate this sad reality: between 1997 and 2002, the number of documents prepared by the EU Council originally drafted in French fell from 47 percent to 18 percent. France is no longer the country that comes up with new ideas. Our experts, diplomats, and officials working in the EU institutions are no longer the ones who take the lead on the drafts that eventually become official documents. People no longer feel it's necessary to negotiate in French because France is less and less likely to play the key role in negotiations.

None of this is inevitable. It is the consequence of our lack of will and imagination. It's because of the resources that we didn't deploy at the right time and place, the French officials we didn't get placed at the right levels in the right institutions, our negligence in transposing EU law into national law, and the low level of interest we show for all these things. By living off our past, by believing that we can get away with anything because we're France, by thinking that we don't have to try as hard as the others do, we are losing influence. It is possible to change.

I want to come back to this outmoded idea that European issues are "foreign affairs" and that the European Affairs Ministry should be part of the Foreign Ministry. European issues have become national issues given how central a role Europe plays in our daily affairs. I think the prime minister should be in charge of European affairs. His interministerial responsibilities would be a great asset. His political weight would reinforce France's influence in the European debate, where we owe it to ourselves to be more active. I would add that I have always found strange the division of labor between the president of the Republic, who's supposed to be responsible for the world, and the prime minister, who is responsible for France but who is not allowed to deal with Europe. By getting involved in European affairs, the prime minister could spare the president from having to attend countless EU meetings. That way the president could focus on summits of heads of state and government where the big decisions are made, leadership is provided, and deals are cut. The president should provide the direction of European policy, and the prime minister should help implement it.

OUR WORLD DOES NOT HAVE TO DISAPPEAR: THE CASE OF ALSTOM

A few months before the presidential election of 2002, Prime Minister Lionel Jospin was challenged by workers of Lu, a company that makes cookies, whose jobs were to be cut by a social plan. Their exchange was tense, and it significantly contributed to Jospin's failure in 2002. The protesters were essentially saying "Shouldn't we vote for the bosses, because they're the ones who are really in charge?" This comment echoed an earlier remark Jospin had made on television, when he was asked about job cuts at Michelin and said pithily, accurately, and pathetically: "The State can't do everything."

In truth, European integration and globalization notwithstanding, the destiny of nations and the evolution of the world still depend very much on states. Globalization creates a new context and different challenges. China and India threaten parts of our economy. But it's up to us to stay in the game. Our world is not destined to disappear. What we have to do is help to find a new balance. This is what led me to intervene strongly in the Alstom affair.

Alstom is a big French energy and transportation company. It's the world leader in the construction of electrical power stations, turbines, and high-speed trains. It's the second-largest maker of cruise ships, suburban trains, subways, and tramways. It has 18 percent of the global rail transportation market and employs 69,000 people, including 25,000 in France. Despite this major industrial potential, in the summer of 2003 Alstom found itself in a difficult financial situation. Given the long delays between the time matériel is ordered and the time the order is filled, this kind of company can function only if the customers pay periodic installments as the work progresses. The installments are in any case backed by banks, so that the customers can get their money back in case the company doesn't produce.

In July 2003, the banks refused to continue to provide guarantees for Alstom because the company's financial situation had gotten worse, through previous management errors. In September 2003, after a tough battle with EU authorities, the French government got exceptional and temporary authorization from Brussels to buy part of the company and to support it financially so that it could restructure. To protect free competition, state aid to private companies is in principle prohibited by EU law. It is, however, sometimes allowed by the EU Commission in special, worthy cases.

In spring 2004, it seemed that the time granted by the Commission would not be enough to save Alstom, and that a new rescue plan would have to be put in place. With the Commission having made clear that it would not extend its authorization because doing so would interfere with free competition, we had two choices. The first, supported by the banks, consisted of merging Alstom with the nuclear energy firm Areva, with the high risk that the problems of the former would spread to the latter. The second, proposed by Siemens, Alstom's German competitor, was for Siemens to take over Alstom's most profitable activities and let the others recover on their own. In reality this was not very realistic.

As for the Finance Ministry, yet again, it genuinely thought there was nothing that could be done. The ministry accepted that the company was finished and that there was no point in resisting fate. A final memo definitively stating this conclusion reached my desk. I did not fail to ask the young and brilliant drafter of the memo to redo it, this time taking the time to imagine what he would have written if his own father had worked for Alstom. I was convinced that we couldn't just let the company's 25,000 jobs in France disappear.

At the highest levels of the French state, the temptation was to promote a Franco-German merger. I agreed, so long as that would not lead to the dismantling of the company. But just one meeting with the CEO of Siemens, Heinrich von Pierer, was enough to convince me that all he wanted was to see Siemens's competition disappear, and there was no point in further discussions. A huge controversy rose up in Germany about my supposed "nationalism." To be honest, I didn't care.

After having looked at all the options, when everything might have led me to give up, I decided to go back to Brussels. After three sessions negotiating with Mario Monti, then European commissioner for competition, I got the Commission to agree to a four-year moratorium and permission for the State to invest new public money in the company. In exchange, Alstom agreed to reach industrial partnerships during this four-year period and to give up some activities representing around 10 percent of its turnover. But I did not accept the closing of a single production site in France or the sell-off of any strategic assets. In parallel, I had to convince the banks to participate in the recapitalization of the company and to provide guarantees for Alstom's customers.

The discussions with the EU commissioner were telling. Mario Monti is an intelligent, French-speaking, and honest Italian. His natural rigidity was nonetheless reinforced by the feeling of omnipotence that came from the Brussels bureaucracy. A commissioner doesn't negotiate alone. He is constantly surrounded by seven or eight colleagues who each represent a different directorate. So his room to maneuver is extremely limited. The bureaucracy had decided that Alstom had to "pay a price." They felt it had been given too much aid, and that it didn't have many months left. I fought this line every step of the way, arguing that the transportation and energy markets were promising. I also believed so strongly in the future of Alstom that I committed the State to invest up to 20 percent of its capital.

Since the agreement reached with the Commission in spring 2004, the value of Alstom's stock has tripled, as has, therefore, the initial investment of the State. What I obtained for 700 million euros twenty months previously the State was able to sell for €2 billion to the Bouygues group. The operation turned out to be a winner for the State, the company, and its workers. Thousands of jobs were saved and a French industrial jewel was preserved.

During these long weeks of negotiation, I went several times to Alstom's industrial sites, particularly in April 2004 when I still had no solution to propose. This was a big political risk. I don't regret it. It allowed me to see just how low the workers' salaries were in this industry. One thousand two hundred euros per month after twenty-five years in the business for people who know how to make high-speed trains is much too little. This observation made me determined to save Alstom, as if it were a shared destiny with the individuals I had met during these visits, who shared with me their passion for their jobs and their pride in their know-how.

Alstom is not Lu, and Lu is not Alstom. And anyway, my complaint about Lionel Jospin is not so much that he didn't save the jobs threatened at Lu or Michelin but rather that he said that saving them wasn't the job of the State. Because after all, political will can sometimes make it possible to preserve jobs and it can lead to profitable investment. Political will is the most necessary quality to renew politics. And even if you believe that "the State can't do everything," it can at least not put our companies in the worst possible position in international markets. When you destroy thousands of jobs with this ridiculously inflexible policy of a thirty-five-hour workweek, and when you prevent France's modernization by refusing to see the world as it is, maybe you can't do everything, but you can at least take responsibility for what you have done.

THE ROLE OF MINISTERS

There is no doubt that part of the success we had with Alstom had to do with the fact that I was a major figure in the government. If

I didn't have this political status, Brussels wouldn't even have listened to me, and it would have gone behind my back to the prime minister and president to get me to accept the idea of a merger with Areva, which had the support of the banks. On the issue of prices in the retail sector, our opponents tried everything to undermine the negotiation, including going to the president's advisers.

In my view of public policy, ministers are not mere colleagues of the president of the Republic. They have to direct their ministries with a large degree of autonomy, in the framework set out by the leader of the majority. The majority leader is the one who should determine the objectives and the timetable. The minister should be involved and should figure out how to accomplish the goals. In this way the minister has the freedom he or she needs to act. If the minister fails, he or she should bear responsibility by resigning—or by "being resigned."

As interior minister, I announced without telling either the president or the prime minister in advance that I wanted to eliminate double punishment from our legal system. I immediately won unanimous support from members of Parliament. If I had talked about it in advance, some technical adviser would doubtless have prohibited me from acting or at least tried to stop me.

Moreover, ministers should not be weakened by changing them every six months, which has been the case for the Finance Ministry over the past few years. Nor should they be flanked by a chief of staff guided by remote control from the prime minister's office or the Elysée. If the minister is so weak that he needs supervision, then he shouldn't be minister.

SEEING THE LONG TERM

The turnaround of Alstom is also proof of the importance, in politics, of knowing where you want to go. I went to bat for Alstom, and I succeeded, because I had confidence in the talents of the women and men who work in this company, because I fundamentally believed in the future of Alstom's businesses, and because I was convinced that industry had to continue to play a big role in Europe.

I don't think that services alone are the future of European economies. I think we've got to maintain industrial activities in a number of strategic sectors. These include transportation and energy, both because we have businesses with global capabilities in these areas, and because these are two indispensable parts of our economic and social life. It was a great mistake to have abandoned steel in the early 1980s. Another mistake was to have allowed Pechiney, the crown jewel of our chemical industry, to disappear, to the benefit of its Canadian rival, Alcan. We cannot maintain all industrial sectors, but we must choose those that seem most promising to us, in areas where we have advantages, and invest massively in them. For France, the agricultural industry is obviously one of these.

I believe competition is necessary. I would like to stress, because it is not said often enough, that the main goal of competition policy is the protection of consumers and their buying power. The role of competition policy is to prevent—by preventive or if necessary punitive action—monopolistic companies from taking advantage of consumers by unduly raising prices or by using reprehensible business practices.

At the same time, competition policy must not prevent the formation of large French or European companies capable of ensuring our long-term industrial presence in strategic, high-tech sectors. A blindly dogmatic and restrictive policy on state aid could have the effect of preventing the emergence of powerful European companies. It would have been crazy to let Alstom die when urban transport, railways, and energy are the key sectors for the future of industry. That's what I explained to Mario Monti, and that's why he listened to me.

It is not illegitimate for the finance minister to promote the creation of national and European champions. I followed this strategy by strongly encouraging Sanofi and Aventis to merge to form the third-largest pharmaceutical company in the world. Since drugs are reimbursed by welfare, we had to maintain a global French pharmaceutical industry, which would not have been the case had Novartis, a Swiss company, been allowed to buy Aventis. Some called me "interventionist" because of this. I accept this label if it means avoiding substantial industrial losses. And I accept it if it means that France can keep control of part of its pharmaceutical supplies, which is essential when health crises such as avian flu or SARS can emerge at any time. The finance minister doesn't have to be a mere spectator.

Serving as interior and then finance minister reinforced my conviction that nothing is inevitable. That doesn't mean it's easy to wield power. It's exhausting. But it's simply wrong to claim that nothing is possible anymore. If we don't do anything we have to accept what happens. But if we act, we have a chance to improve things. Are we going to allow France to just sit there and passively watch its transformation and the success of the countries that surround it? Why shouldn't we also enter tomorrow's world holding all our trump cards?

CHAPTER 5

Society and Schools

THE CRISIS OF THE SUBURBS

The crisis in the suburbs in fall 2005—when the accidental deaths of two Arab youths near Paris led to several weeks of riots and protests—will always remain important for France. It was a true moment of awakening for me personally. The reactions to and countless commentaries about these events were so characteristic of the malfunctioning of our democracy. Conventional wisdom has rarely been as widely shared. Thus it was said that the riots were mostly a "social" protest; that most of the rioters were primarily victims; and that the guiltiest party of all was the State broadly defined, because it hadn't done enough, spent enough, or provided enough education, training, or assistance.

Once again, society as a whole was criticized in a search for collective responsibility. The only goal seemed to be to avoid in advance any sort of individual responsibility, and that was indeed the effect. For as always in France, since everyone was responsible, no one was guilty—and no one had to admit his shortcomings. The usual explanation was offered: if something didn't work, that's because we didn't spend enough. And the only possible response is to spend more, probably to get even less! This endless refrain at least has the benefit of being useful over and over again—for the suburbs, but also for integration, for education, for lack of opportunity, and for training. The worst thing is that most of these commentators believed what they were saying.

During the crisis, the conventional wisdom was propagated by some famous people who originally came from troubled neighborhoods and sought to become spokespeople for those who now lived there. They hadn't themselves lived there for years, but because of their success in show business they felt they should step forward. The locals were more surprised than anyone by the sudden expressions of concern. Thus we saw the strange spectacle of rapper Joey Starr pretending to be a big shot and comedian Jamel Debbouze acting as an emcee at discussion forums. We even saw former tennis player—and now singer—Yannick Noah explaining that he would leave France if I came to power, failing to acknowledge that he hadn't lived in France for years anyway. For all these public figures, the explanation was clear: there was only one guilty party, and it was me. Perhaps there was another: the police. The solution was thus obvious: the interior minister had to be fired and the police had to withdraw. Thus would calm be restored, and the suburbs would live happily ever after.

The problem is that behind all this banality, behind the compassionate and exculpatory rhetoric, an increasingly serious situation was developing and deepening. Since the early 1980s, France had spent billions in the suburbs. It put dozens of successive plans in place. Not only did nothing change, but the situation got worse. It wasn't money that the suburbs needed, but new policies, different methods, and more honesty.

What our neighborhoods really need is regulated immigration. Without this, nothing will be possible. This is a disturbing reality, but it's the truth. Many of the current problems of our suburbs are the result of unchecked, and therefore poorly managed, immigration. The paradox is that the children and grandchildren of the first generation of immigrants feel less French than their parents and grandparents, even though legally they are French. Saying this leaves you open to caricature. But I take that risk because this is the reality.

SEMANTICS

I want to come back to the context in which I used the word *ra-caille*—scum—one evening, on the promenade in Argenteuil, a northwest suburb of Paris. I wanted to go to this part of town that was known to be one of the most crime-ridden in the whole Paris region. I chose to go at night to show the local thugs that henceforth the police would be welcome everywhere and at any time. I came to set up a new Security Company of the Republic—CRS, the French anti-riot force—reinforced by its new operational doctrine. In 2002, I had changed the CRS's operational doctrine—the missions they are given and the ways in which they carry those missions out. Rather than having them constantly deploying all over France, depending on events, with a lot of time wasted moving around at great cost to the taxpayer, I decided that the CRS should henceforth have regional assignments. This would reduce the financial costs and improve the family life of the CRS agents. In particular, this new operational doctrine would allow agents to get to know their neighborhoods better, which is indispensable if you want to go after drug dealers and dismantle gangs, and not just maintain the status quo. People should be able to live in peace in their neighborhoods. Young women should be respected. Working at school should be more attractive than being a lookout for a drug dealer. The State should take an interest in the sources of revenue of people who don't do anything all day but somehow drive a Mercedes. Withdrawing the police from these neighborhoods is the exact opposite of what is necessary. The law of the Republic should be enforced in the neighborhoods.

When we arrived in Argenteuil—and it wasn't by coincidence—two hundred angry people were waiting for us. They were vigorously insulting us and throwing everything at us they could get their hands on. The air was thick with tension. The security forces were on edge. I decided nonetheless that we'd do the final four hundred meters on foot. It wasn't exactly a walk in the park. I didn't want our procession to rush. The hooligans went all the more crazy when they felt there was a provocation. This was their turf.

They saw my very presence as a challenge. What an inversion of values!

What perverted thinking! The law of the gangs standing up to the law of the Republic! The brawl was violent and lasted almost an hour. I waited at the Argenteuil police station while the CRS secured the area. Around midnight, I was able to continue my visit. As I found myself at the foot of some tall towers, a window opened up and a woman who looked to be of North African origin called out to me: "Mr. Sarkozy, get rid of this scum for us! We can't take it anymore!" I responded: "Yes, madame, that's what I'm here for. I'm going to get rid of this scum for you." Neither she nor I realized at the time we'd be such a big hit.

Despicably, some who saw themselves as the conscience of polite society grabbed on to the word and started accusing me of anything and everything. In twenty-four hours, I was said to have insulted young people, encouraged racism and xenophobia, lost my nerve, and much more besides. To hear the left tell it, it was the very use of this word that set the suburbs on fire. Forget the political machinations, which you could argue were fair game. Much more worrisome was that some of our elites took this analysis at face value. That says a lot about the gap that exists between what those who live in the suburbs think and what is said by those who are far away. Indeed, the farther away they are the more they talk. I'll bet that these neighborhoods have never been more talked about than at symposiums and dinner parties in late 2005.

That's the hard part: to resist the pressure of conventional wisdom without being caricatured as excessive or accused of holding unacceptable positions. I never thought that I went too far in using the word "scum." I described a situation that I hated, one in which gang law prevailed and thousands of our fellow citizens lived in fear. I called some individuals that I refuse to call "youth" by the name they deserve. I don't think they should be confused with the young people who have nothing to do with this violent minority. Similarly, I am not afraid of saying that those we call "big brothers" are more often crime bosses and gang leaders than they are examples of success through work and merit. And finally, I never understood in what way this common expression should be seen as stigmatizing skin color, which I know perfectly well does not predispose anyone to become a delinquent. I abhor racism and I hate xenophobia. I believe in the power and wealth of diversity. I love the idea of a France that has many sides. But I accuse those who deny the reality of the lives of our least advantaged citizens of being the real cause of the rise in extremism, because they have condemned the Republic to blindness, passivity, and immobility.

Then there is the issue of what kinds of words ministers should use. For example, the spokesman of the Socialist Party admitted that the word "scum" was common, but he insisted it was not appropriate coming from the mouth of a "ministerial excellency." This is a strange conception of the Republic. We're all supposed to be equal in duties and rights. There is thus not, in my mind, one set of words for elites and another for the people. There is honesty and dishonesty. There is straight talk and hypocrisy. There is vulgarity and proper language. I never felt that by saying "scum" I was being vulgar, hypocritical, or insincere.

A STERILE DEMOCRATIC DEBATE

A political leader must make him- or herself understood. To that end, the best approach is to speak simply but not simplistically. The goal must be to be heard and understood. The hard part is to achieve these goals without lowering the level of debate. The controversy around my use of the word "scum" was sterile and unimportant, but it was revealing about how some of our elites behave—so politically correct when it comes to style, so conservative when it comes to substance. This is the tendency I want to break, and I have no qualms about that. It's this sterile conception of the democratic debate that is responsible for the public's widespread boredom with politics these last few years. This is serious business. You get in trouble if you seek to break taboos, violate codes of behavior, or take the risk of innovating. Even before people have taken the time to think about the relevance of your proposal, they're accusing you of "populism."

If you fail to use the usual language, propose an original idea, or even simply bring attention to something citizens are worried about, you're immediately called a demagogue by a good portion of the intelligentsia. Even worse, whoever dares to take such risks can be accused of "trying to compete with the National Front." I've had this experience several times.

In June 2005, after a woman was murdered while jogging, leaving behind an eleven-year-old child, I raised the question of the responsibility of the magistrates who had released the suspected killer—Patrick Gateau—from prison. We're talking about a repeat offender who fifteen years previously had been sentenced to life in prison for a similar crime. How can you explain to a weeping husband and a devastated little girl that the State allowed such a monster to move in next door? Being a judge is obviously a difficult job, which must be done in a strict legal framework. No judge could claim that he or she had never made a mistake. But when society gives magistrates the power to make a decision as fateful as that of early release of a repeat offender, to increase that person's chances of reintegration into society, it can expect certain things in return. It is legitimate to insist that every precaution has been taken to be sure that the decision does not put others at risk.

Let me be clear: if letting someone out on probation presents a risk, the potential victim shouldn't be the one to run that risk. I would rather err on the side of reducing a repeat offender's chances of reintegration than to err by exposing an innocent victim to a criminal's deadly impulses. If you demand that everyone take responsibility for his or her methods and decisions, you lower the risk of error and you improve the way things work. This is what has happened with doctors, elected officials, youth leaders, sports figures, and all those who play a role in maintaining the security of our citizens and whose professional responsibility is now questioned more often and more rigorously than before. These new demands have led them to improve their procedures and change their methods so that everyone's life—that which is most precious to all of us—is better protected.

In a normal democracy, it would have been natural for the magistrates who set Gateau free to be called on to explain their decision, and possibly to be punished. In French law, this is difficult if not impossible, and in any case it never happens because political leaders fear the reaction of certain magistrate organizations. In case of a disciplinary error, only the justice minister and the first presidents of the Court of Appeals can go to the Superior Council of the Magistracy, which they rarely ever do. When the State has been ordered to pay reparations for failing to provide public justice, the minister can also go after the guilty magistrates. This is what is called a "recourse action." It's never been used, and that is frankly a scandal.

There was a huge uproar about my remarks on the Gateau affair. I was accused of "talking like the National Front," and someone close to the president of the Republic went so far as to say that "suspicion of the magistracy is the beginning of social breakdown." I think that social breakdown begins when people have power over the lives of their fellow citizens but don't have to answer for anything.

I'll never forget being in the little church of Seine-et-Marne in the summer of 2005. It was very hot, and the crowd was growing. There was great emotion. I admired the husband who did not let his emotion show, and the little girl whose distress was written all over her face. I apologized to them on behalf of the State. But I'm not sure they were listening to me at that moment.

Conventional wisdom focused on the isolated nature of this case and the impossibility of eliminating risk entirely. Unfortunately, however, the Gateau affair is not an isolated case. Before that, there was the Gregory affair and two broken families—one because their young son died, the other because a member of the family was prematurely accused and murdered by the father of the child. There was the Mourmelon affair, in which the disappearances of several conscripts of the military contingent deployed in that city were erroneously considered to be desertions for years. There were the Fourniret and Bodein affairs, in which monsters were allowed to live next door to decent people. There was the Jean-Luc Blanche affair, in which a serial rapist on probation was interrogated for a sex offense against a minor but then set free by an overworked judge—like so many judges—before going on to rape no fewer than four women in the summer of 2003. There was the Dils affair, in which a sixteen-year-old boy was sent to prison for fifteen years for two crimes he wasn't guilty of. And unfortunately there have been many other examples. I have met enough victims and families to understand the deep suffering that results when the justice system fails to do its job in protecting the innocent. In most of these cases, the State was made to pay reparations, but the magistrates in charge of these matters never had to testify, provide explanations, or take responsibility for their actions. I want to put an end to the lack of accountability that separates the French from their justice system.

In the case of the disappearances at Mourmelon, it was by chance that Pierre Chanal, eight years after the first disappearance, was caught red-handed and arrested for the kidnapping of a Hungarian hitchhiker. He was convicted for this crime in 1990 and freed in 1995. Between 1995 and 2003, he remained free while waiting to go on trial for the disappearances. The day after that trial started, he committed suicide in prison, leaving the families to wonder forever about the motives and circumstances of these tragedies and suffering the pain of unpunished crimes.

In the case of the disappearances in Yonne, the justice system for years deemed it credible that seven young, handicapped women, all with similar profiles, all ran away from home in the same region in less than six years. Numerous pieces of evidence were lost or misfiled, wasting a huge amount of time. For various reasons of substance and procedure, the only penalty paid in this matter was the nonpayment of an honorarium for an already retired magistrate.

Today, after the legal catastrophe of the Outreau pedophile case, no one disputes the need for a new system for holding magistrates responsible for their actions. Such a system must of course take the difficulties of this job into account, but it must allow society to ask magistrates to account for their actions and to avoid the discredit that the negligence of a few casts on an entire profession. You can't have power without responsibility. I regret to have to make this point so strongly in order to try to get it across.

THE NEED FOR SOCIETY TO PROTECT ITS OWN

When I started to act in the area of prevention and punishment for sex crimes, I was similarly attacked. Many sex offenders are unfortunately incurable. The risk of repeat offenses is very high. This is an established scientific fact.

Society must protect itself against people whose sickness transforms them into predators.

As interior minister, I had my most painful moments in my meetings with victims' families, including parents of murdered children. I hesitate to invoke my own distress, since it's obviously trivial compared with the suffering of those whose lives will no longer have any meaning. I hate the way these crimes are sometimes listed as "news items" in the newspaper. A murdered child is not a "news item" but rather a tragedy and a failure that demands our response. The State must do everything it can to avoid such unspeakable scandals.

My last visit was to the family of young Mathias, a four-and-a-half-year-old boy, who was raped and drowned by a deviant. He came from a happy family of farmers, living in the Nièvre. Here, too, I will never forget the father who was waiting for me on the farm's doorstep: "Am I receiving the minister or the man?" he said to me before any other greeting. "It's the man, the father," I responded in a voice that revealed all my emotion. "Well, there you are," he continued. "It's my birthday in two days. Nice present, my son raped and killed!" How to respond? What to say? What to do? Probably nothing, other than simply be there to help bear an inhuman burden.

Once inside the house, I hugged Mathias's mother, a young woman of exemplary dignity, who held back her tears without hiding the state of tremendous confusion she was in. They suggested I take a place on the sofa, in the middle of which was Mathias's teddy bear, all alone and sad. I had tears in my eyes. We didn't say much to one another, but our silence was enough. Why Mathias? Why was this monster there? Why not reestablish the death penalty? This was the question that the father kept repeating. Understandably! I did not have the heart to tell him that the death penalty had no deterrent effect whatever for insane people and deviants. This is why my personal philosophy led me a long time ago to oppose it.

I was aware that "my philosophy" didn't matter very much to parents suffering the loss of a martyred child. I often think about this family and what it has gone through.

Death penalty or no death penalty—this debate comes up every time a child has been killed in such a monstrous crime. But there are other answers. Keeping a database of sex offenders is part of the answer. And I had to break a lot of taboos, and fight every step of the way against lies and rationalizations, to get such a database created in 2004.

None of the existing databases we had actually did what I wanted the database of sex offenders to do. We kept legal records listing indictments, a database for digital fingerprints and DNA samples, and the police and gendarmerie records that kept track of criminals and the facts and circumstances of their crimes, but none of these kept up-to-date addresses of the people in question—this just wasn't their purpose. The database of sex offenders is preventive. Its purpose is to allow us to know the address at any given moment of anyone who has been convicted of a sexual offense. It requires all offenders to notify the authorities anytime they change addresses, and for those guilty of the most serious crimes to check in monthly with the local police station or gendarmerie. Thanks to the existence of this sort of database, when a child is reported missing in Canada the police can immediately go to the homes of those on the list who live in the child's neighborhood. Here, too, the evidence is quite clear: you have to act in the first few hours of a child's kidnapping before it turns into an irreversible tragedy.

When I announced the creation of this database, I was accused of violating human rights—my goodness!—and the National Consultative Commission on Human Rights denounced an "excessive infringement on private life and the right to privacy" of the convicted. This makes me think of the hamlet in Moulins-Engilbert where the parents of little Mathias live. I remember his little bike still parked in the courtyard and I wonder about this revered "right to privacy." In the view of the National Consultative Commission on Human Rights, who should benefit from this right? This repeat offender suspected of having martyred this child or this family whose life stopped on the weekend of May 8? Where was Mathias's family's right to privacy? A State governed by the rule of law must find the right balance between protecting victims and proportional punishment. But there are certain expressions, such as "right to privacy," that are borderline offensive. Someone who has violated a child does not have a right to privacy. Such people have the duty to never do it again and to have themselves put away by the law and society so that they can never hurt again.

I want us to go even further in the area of sex offenses. After serving their prison terms, sex offenders should receive psychological and psychiatric treatment, and the police must be able to track the most dangerous offenders with an electronic bracelet. The legal system should be able to make this mandatory. If the law I have put forward is approved, this will become legally possible. However, the Constitutional Council has already ruled that the principle of nonretroactivity in our criminal law means that we may not put these measures in place for offenders who have already been convicted, even though there is no question about how dangerous they are. Immobilized by this principle, the thinking of our experts, elites, and magistrates remains stuck, as if behind a brick wall they can't see beyond. Personally, I'm not afraid of saying that a principle, even a constitutional one, can and must be amended if its effect could be to do physical damage to honest people, especially children. I'm only suggesting that we take precautionary security measures. This has nothing to do with the principle of retro-activity of the criminal code. Besides, such measures have been put in place by most of our neighbors, who share the same founding principles in the area of criminal law as we do.

The reactions of the chattering classes are sometimes so extreme that they end up discouraging people of goodwill from trying to innovate. Many end up thinking that it's better to be wrong with others than right alone or almost alone. Having understood that, I decided not to speak with political commentators in mind anymore, but to speak for the public itself. And I've never regretted it.

THE DIFFERENCE BETWEEN BEING POPULAR AND BEING POPULIST

I want to talk about the difference between populism and popularity. Being popular is talking about things that matter to the French. Being popular is being understood by your fellow citizens. Being popular is being more offended by a problem than by the proposed solution. Being popular is trying to improve people's daily life. Being popular is refusing to abide by convention.

Being populist, on the other hand, is thinking that an opinion is true simply because it's widespread. Being populist is thinking that elites and civil society leaders are never allowed to speak for the people. Being populist is looking for popular support without doing what is needed to resolve a crisis. It's chanting slogans without undertaking reforms. It's talking about everything without proposing anything. This is not my style.

I've never been afraid of taking unpopular positions, for example on affirmative action. I do not wish to see the disappearance of elites, civil society institutions, or regulatory and representative bodies. Thus I put all my energy into the creation of the French Council of the Muslim Religion. I'm not a big fan of referenda for deciding on economic and social reforms. They run the risk of reducing complex issues to simple, binary choices. I believe in representative democracy. Similarly, I would like to see France relearn the practice of social dialogue, which means promoting the emergence of stronger and more representative unions. For this we need to end the monopoly that the five main union federations have on presenting candidates in the first round of elections of labor representatives, which has been the practice since the Second World War. A law should give union and industry representatives deadlines for resolving issues such as workers' rights, unemployment insurance, and retirement, at the end of which the government and Parliament would get involved in case of an impasse.

But I would like to see elites and civil society wake up, open their eyes to the realities of French society, and rediscover a love for thinking and daring to think. The wall of conventional wisdom blocks as many good ideas as does a failure to think at all.

HAVING YOUR IDEAS CARICATURED

I am convinced that our democracy suffers more from a lack of debate and criticism than from too much. This belief led me to take a position unambiguously in favor of the cartoonists who created a scandal with their drawings of the Prophet Muhammad. I can hardly be suspected of bias in favor of cartoonists, who, I think it's fair to say, have not exactly treated me kindly. I've been caricatured in every way on every subject. My private life, my appearance, my words, and my politics have all been covered. They've tackled everything, and not always very elegantly. It often hurt.

But, however excessive it can be, caricature is useful for democracy. It obliges leaders to keep their feet on the ground. It often sums up current affairs or someone's temperament in a useful way. It represents an area of freedom that democracy needs. There should not be any taboo subjects—otherwise there would be a very long list of them. I believe in God, and I sometimes go to church, but I believe that religion, like power, must accept criticism, caricatures, and mockery. This is true for all religions, including the one most recently arrived in France, Islam, which cannot claim equal rights with the others without accepting equal duties. What would be disrespectful to Muslims in the end is not a few caricatures mocking the Prophet just as they mock Jesus Christ. What would be disrespectful would be to consider France's Muslims different kinds of citizens from the others.

Since 2002, my ideas have often been caricatured. For example, I call on State authorities to do more to integrate young people of immigrant backgrounds, so that ethnic communities don't turn inward because of the State's failure, and I'm accused of promoting ethnic divisions.

I propose for the first time in years a tailored immigration policy, meaning one that explicitly recognizes the benefits of openness for a country such as France, and people accuse me of encouraging extreme nationalism.

I comment that no one is required to remain in France, that when people are welcomed somewhere they must respect and if possible love those who welcome them, and I'm accused of xenophobia.

I call for violent behavior among young people to be detected and dealt with as early as possible, and people accuse me of turning three-year-olds into criminals.

It's much simpler than that. To take just this example, everyone knows that during recess at school some children are abnormally violent at an increasingly young age. No parent or teacher could seriously claim that he or she does not know the difference between a child who is alive, petulant, communicative, or even boisterous, and a child whose only form of self-expression is to hit classmates or even teachers. I have enough common sense to know that an abnormally violent child will not necessarily be a criminal. I never proposed to keep a list of these kids. What I do believe, on the other hand, is that a violent three-year-old is a child who should be placed into care. The only chance of being effective is to act at as early a stage as possible. It is necessary to understand the reasons why the child is suffering, whether it's because he or she has been mistreated at home or is just going through a difficult period. It's the duty of society, the education system, the parents, and the school health service to help the child, and for that you have got to identify the problem and deal with it. It is an established fact that untreated suffering can later lead to criminal behavior. A number of criminals, particularly sex offenders, were themselves beaten or raped as children, victims before becoming tormentors. The gang of savages who tortured and killed Ilan Halimi in February 2006 were already known for their violent behavior when they were fifteen-year-olds at school. Who sought to understand this violence, who tried to speak to them, to offer them an answer that might have avoided the spiral into barbarism? Sadly, no one. I don't know if my ideas are all good ones, but I do know that the current situation is all wrong.

Fifty years ago, the school health service did a good job of looking after the weight, height, eyesight, and hearing of the pupils. Today, when the great majority of children have a family doctor, we expect more. The school health service must get fully involved in prevention, the poor cousin of our public health policies. It should deal with prevention of obesity, prevention of addictive behavior, and prevention of the risk of excessive exposure to the sun. It should provide information to children about good health care practices such as regular visits to the dentist and to a general practitioner. It should promote physical activity, and much more. We have so much to gain from this: lower health care expenses and better health for our fellow citizens. And this same health care service should identify, care for, evaluate, and follow up on behavioral problems, as much to prevent these problems from becoming incurable as to help prevent adolescent suicide, the rate of which is tragically high in France. Every day an increasingly painful incident is reported that should make us think about this situation. In Evry, one sixteen-year-old was murdered by another one. Two broken lives. Why?

Some worry about the risk of stigmatization. I don't understand this argument. Everyone sees it: the more violence, the more young people are fascinated by this violence. We can't just stand there and do nothing. We run no risks by taking action, but we run many risks by continuing to act as if there were no problem.

I am not sure that people realize how much it takes for me to clarify, correct, convince, and try in the end to make progress. But if I am the subject of so much caricature it's because I attack so many sacred cows. Caricature doesn't bother me if ultimately French society agrees to move forward. That's the goal: getting France moving again.

PRIORITY EDUCATION ZONES

The issue of Priority Education Zones (ZEPs) is another good illustration of how difficult it is for conventional wisdom to react other than with indignation whenever someone tries to think and act differently.

At the end of November 2005, not long in fact after the crisis of the suburbs, I declared that it was necessary to "draw up the balance sheet" for the ZEPs. This was at the end of my party's convention on injustice. The expression was admittedly blunt. But at least it was clear. It created an uproar on the left and on the right. People accused me of wanting to cut the extra financial support that ZEPs received, even though I said nothing of the sort. They accused me of wanting to give less to those who already have so little, as if I were so thick that I wanted to make elite schools such as Henri IV or Louis-le-Grand the priorities for the State education system.

People accused me of failing to appreciate the role of education and training in promoting equal opportunities and social advancement. The opposite is the case. I think that one of the main reasons for the failure of the ZEPs is that standards were reduced when they should have been raised. Children of professionals or teachers have everything they need in their family environment to develop a good intellectual and cultural base. But it's at school that the child from a disadvantaged social milieu can have a chance to get to know our great authors and philosophers, to become aware of the importance of history, to discover that after the great effort that's necessary to do demanding science, there is also pleasure. It is this degree of discipline, with all pupils—girls and boys, kids of doctors or farmers or workers—that made the teachers of the Third Republic a success. They knew that by not taking the easy route but by keeping high standards, you got the best out of childhood.

Finally, and most scandalously, by attacking the ZEPs, I allegedly denigrated "the thousands of admirably devoted teachers who do a remarkable job in these schools."

Unfortunately, the facts are undeniable: the ZEP policy has failed. It was created in 1982 and supposed to last for four years. Twenty-three years later, there are more than seven hundred ZEPs. The students' achievement level is much lower than in other schools. The difference in performance does not result only from the fact that the kids in the ZEPs on average come from more difficult backgrounds than the others. In fact, the gap is growing. It goes without saying that the difficulties that the young people from the neighborhoods have are tremendous. The ZEPs have become educational ghettos. Well-informed families, or those with means, avoid them, and as a result the most disadvantaged children are placed all together when what's necessary is to spread them out. Let's be honest: the teachers in the ZEPs are the youngest and least experienced and the turnover rate is much higher than elsewhere. And very few teachers in the ZEPs send their own children to ZEPs. If the ZEPs were the educational success that I was audacious enough to criticize, you'd see teachers staying there and sending their kids there.

To say that doesn't take anything away from the merit of those who teach in the ZEPs. No one is questioning their competence or their dedication. Nor does it take anything away from the success of certain schools, such as the Saint-Ouen-l'Aumône secondary school in Val-d'Oise, which has had great success with its pupils, particularly by cooperating with the Institute for Political Studies (Sciences Po) and the ESSEC business school in Paris. All this success resulted from the great determination and efforts of the teachers

and administrators to change old habits, have the courage to innovate, and find partnerships. All I'm questioning is the way the ZEP
policy has been implemented over the years, nothing more.

The supplementary financial support that has been given
to the ZEPs has been insufficient and spread out over too many
schools. It's almost an insult to spend just 1.2 percent of the national
education budget for such an important priority, which concerns 20
percent of all students. Moreover, this money is almost exclusively
used to reduce the number of pupils in each class—twenty-two per
class compared with twenty-four in non-ZEP schools. This reduction in class size is too uniform and far too small to have an effect
on the success of the students. It's only when you get fewer than fifteen students per class that the reduction of class size starts to make
a difference. The factors that account for success at school are well
known: family environment; living conditions, particularly having one's own room; social diversity; and finally, and especially, the
pedagogical abilities of the teachers. The ZEPs, as they've been run
for the past twenty-five years, don't help on any of these fronts.

My goal is not to get rid of priority education, which is indispensable. Nor, however, do I think we should just keep throwing
more money at it, which will lead to the same failures if we don't
change our approach. Rather than thinking in terms of zones that
stigmatize, rather than concentrating all our means on the same
reduction of class size everywhere, I want us to widen the range
of tools we use and start thinking in terms of individual students,
as in the Netherlands and Sweden. The means devoted to priority
education should serve to give each student who needs it—whether
or not he or she is educated in a ZEP—the appropriate kind of help.
There are lots of good ideas out there. There's early and reinforced
child care between the ages of eighteen months and four years, because this is the age when children develop many of their cognitive
skills. There's educational support, and tutoring. There are boarding schools of excellence, where evenings are quiet. And there are
lots of other ideas.

I also think we should encourage private schools under contract with the State to go into disadvantaged neighborhoods. They're asking to do so. They've got projects. But everything has long been set up to discourage them. Whatever we think of them, private schools are highly valued by families. There are long waiting lists. Parents are reassured by the greater degree of structure and by the greater association with the educational development of their children.

The Socialist program for 2007 proposes to link the means allocated to private schools to the degree of the schools' social diversity. Whether this would be done by giving bonuses to schools that the Socialists deem to be sufficiently diverse or by fining excessively advantaged schools, the result would be to punish those schools that are insufficiently diverse. The same schools that were dissuaded, or even prevented, from moving into disadvantaged areas by the famous 80/20 rule will now be penalized for having done so! The tacit 80/20 rule resulted from the 1984 agreement between the Catholic education system and the then Socialist government. It limits support for private school to 20 percent of total primary and secondary spending on education, and it encourages this division in regions such as Brittany, where private Catholic schools play a historically greater role. The rigid application of this rule at the local level makes no sense: in places where there's high demand for private school, it would deprive parents of the possibility of sending their children to the schools, whereas in other areas, the demand for private schools might not even be enough to fill up all the slots.

In fact, instead of trying to understand the reason for the current success of private schools so that we can apply these lessons to public schools wherever possible, the Socialists prefer to punish private schools. They prefer this to encouraging the market, and letting all French people, in particular the most disadvantaged, have access to private schools if they want it. This has unfortunately become a habit: take what works and destroy it. I propose, on the contrary, that we take what works and make it accessible to everyone.

CHAPTER 6

Rewarding Merit and Work

THE BEST SOCIAL MODEL:
ONE THAT GIVES EVERYONE A JOB

Reforming our famous social system will certainly be one of the hardest taboos to break. But let's hope we're able to do so before it collapses like a house of cards.

We have gotten used to bragging about our social model, to the point of proposing that the rest of the world adopt it. On paper, it's obviously great. The labor code is thick. Workers cannot easily be laid off. Short-term workers are relatively well protected. Social provisions are made for people going through a rough patch. The fiscal system and system of national insurance benefits ensure a good degree of income redistribution. We treat pensioners and sick people generously. Our public services are of high quality, our roads are magnificent, our schools are free, and our health care system is among the best in the world. In short, France is an area of freedom, prosperity, and solidarity that is the envy of the world.

The proof of this, it is often said, is that France attracts a lot of capital from abroad. According to the official mantra, "Corporate taxes are high, but the quality of life is better." Yet I can't help pointing out that 46 percent of the value of companies listed in the CAC 40 is owned by foreign investors, mostly pension funds. Moreover, most of these funds are used to buy French companies, which are then dismantled and sold off elsewhere. After Pechiney was sold to Alcan, most of its decision centers were transferred to Canada. In ten years, nine thousand French companies have come under foreign ownership whereas France has taken over only 650 foreign affiliates. One out of every seven French workers (other than in finance and administration) works for a foreign company, compared with one out of ten for Germany, Britain, or the Netherlands and one out of twenty for the United States. It's obvious that France lacks capital to invest in the domestic economy, and the level of foreign investment in France may be more a sign of our economic weakness than a sign of our attractiveness. In addition, the amount of French capital that leaves France is twice as high as the amount of foreign capital that enters. France is gradually losing substance.

Governments of the left, and sometimes of the right, have systematically conducted policies that discourage the creation or holding of wealth in France, and it's easy to see where these policies lead. Our fiscal policy, which very heavily taxes the most mobile factors of production, such as capital and highly qualified workers, has led to the near-disappearance of French family businesses. This has been great for the Belgians, the Swiss, and the British, who are delighted by the windfall of having the richest French people come to live in their countries. Equality should mean not that we all become poor, but rather that we can all hope to become rich or at least ensure social advancement for our families.

Until very recently, expressing doubts about our social model could get you in trouble. After President Chirac did so in his July 14, 2006, interview, he had to quickly clarify his remarks, insisting that "the French social model is neither ineffective nor outmoded." This is because attacking the French social model is like attacking French national identity. Now French people no longer kid themselves.

Since 1984, a generation ago, the unemployment rate has hovered around 10 percent. It goes up during recessions and falls a bit during periods of growth, but the structural hard core of unemployment does not go away. This is not an inevitability linked to the European economy—many of our European partners have returned to full employment, including Britain, of course, but also the Netherlands and the Scandinavian countries. The best social model is one that creates jobs for everyone, and this is obviously not ours, since our unemployment level is twice as high as that of our main partners. Once again, I'm not trying to be provocative for the sake of it but trying to wake people up.

The risk of unemployment felt by the French is in any case much greater than 10 percent if you add to the official number of unemployed all those who are artificially excluded from the statistics. These include people registered for unemployment benefits at the National Employment Agency (ANPE) in uncounted categories; beneficiaries of the Minimum Integration Income (RMI) not registered for unemployment benefits; senior citizens who don't have to look for a job; and people on special contracts. If you then subtract the public sector employees (5.2 million civil servants and others) who are not at risk of losing their jobs, you get a real unemployment rate of 20 percent. For the real figure, in other words, you have to count the number of unemployed as a share of the market in which they are looking for jobs, which is to say the private market. This is why people are so worried.

It's an interesting paradox. The French have never talked so much about job security. But they talk about it as if it's a risk for some time in the future. Thus they grab on to today's reality and try to preserve it by refusing to change anything. But job security is a problem now, and the only way we're going to deal with it is to change. I'm convinced that a conservative approach is the surest guarantee of maintaining job insecurity. Reform is the only way of creating the security that workers need.

The situation of workers has also deteriorated over the past twenty years. This is not primarily because of the "financial capitalism" Laurent Fabius inveighs against, or because of this "free market globalization" denounced by the crypto-communists and other anti-globalization activists, but simply because with such a high un-employment rate workers are in a weak position vis-à-vis employers. The number of workers with a temporary employment contract is now up to three million, and more than 70 percent of new hires are employed in this way. The lack of flexibility in our labor laws, and constraints on downsizing, lead companies to prefer this kind of hire. It mostly affects women. Women account for 80 percent of part-time workers, 80 percent of temporary workers, and 80 percent of poor workers. Wages and buying power are stagnant. More than half of the French earn less than €1,500 per month. Regular increases in the official minimum wage, while other wages are stagnant, mean that an increasing number of workers are paid the minimum wage. These workers feel that they're climbing down the social ladder. In ten years (1993–2004), the percentage of workers paid the minimum wage has doubled, from 8 percent to 16 percent of workers.

Finally, in France there are 1.1 million children living below the poverty line and 3.5 million people on welfare. The latter fig-ure would be 6 million if all those eligible were included. Thir-teen percent of retired women live below the poverty line, and a further 25 percent are barely above it. Fifty percent of people on welfare remain dependent on it three years later, and 30 percent five years later. Our system provides easily enough for the socially excluded to survive. But it doesn't provide enough to enable them to overcome that exclusion and free themselves from dependence on the State. What a waste! Certain sectors of the population and certain areas are in very difficult situations, which the average un-employment rate doesn't manage to convey. The unemployment rate for unskilled workers is 15 percent. It is more than 20 percent in regions hit by deindustrialization and in what the government has identified as special inner-city zones (ZUSs). It is 22 percent for those under twenty-five and nearly 40 percent for low-skilled youth who live in ZUSs. Chirac based his 1995 campaign on this "social fracture"—enduring poverty and human suffering—and he had political success with that. It's not clear that the proposed remedies were commensurate with the gravity of the situation.

New inequalities are emerging among the French in terms of the risks they face and their relative confidence in the future. Some people can hope to own property, and others cannot. Some face a high risk of unemployment, and others are sheltered from this risk. Some have long-term job security, and others go from short-term job to short-term job. Some have a high-level degree that essentially guarantees them a job for life, and others didn't get the opportunity at the right time and will probably never be able to catch up. Some are well informed enough and have good enough contacts to get into the best schools, go to the best hospitals, or get public housing, and others don't have good enough contacts to get by in our stagnant society. And this is not even to mention the gap between those of our fellow citizens who still have hope and those who don't.

Beyond this mediocre social record, the French also know that our public services, health care system, and social security system are all sitting on a financial time bomb—our public debt. The 2004 reform of the health insurance program helped to limit spending on health care, but for how long? Those responsible for this system still haven't found the means to guarantee its long-term financial footing. Yet health care spending is certain to keep rising because of our aging population and technological progress. And this is not such a bad thing after all. In the future, we'll have advanced—and spectacularly costly—treatments for cancer, genetic diseases, and neurological problems. This provides great hope. But will we do what is necessary to make sure the money is there so that they benefit all of the French? This is a major challenge, a real test of our solidarity.

By selling off all or part of its holdings in public companies, the State has earned €82 billion (not accounting for inflation) since 1986 and, depending on the estimates, it will earn some €110 billion to €125 billion more with future privatizations, including that of Eléctricité de France (EDF). Even by selling all the "family jewels," we'll thus still be very far from having the amount of money necessary to pay off a debt of €1.1 trillion. Only economic growth can get us out of this.

I think, in any case, that there are much better ways to use this money. The State urgently needs investment in its restructuring and modernization. Yet everyone knows that far from raising money, these things cost money. But no industry or business ever managed to restructure without investing a lot of money in advance. Why would it be different for the State? I see no reason why we shouldn't create a specific budget only for ministries that undertake major restructuring programs. Reductions in operational costs would be compensated with credits for investment. The budget could be augmented by revenue from future privatizations. Using profits from the privatization of big public companies for the modernization of the State rather than for operational expenses would be a good thing. The budget could be managed directly by the secretary general of the Elysée. I don't think you can ask the prime minister to undertake governmental reform. This would be like asking him to cut off his own arm.

To sum up, France is like the Anglo-Saxon countries when it comes to inequality and poverty—but without their social mobility and full employment—and like the Scandinavian countries when it comes to public spending and taxes, but with greater unemployment and deficits. We have all the disadvantages of the two systems without the advantages of either. France's leaders say the French social model is still effective. But the French people no longer believe them.

THE NEED TO CHALLENGE OURSELVES

The French have never before spoken so much about justice while allowing so much injustice to prevail. Never have French leaders so frequently sprinkled their speeches, plans, and laws with the word "social"—social cohesiveness, social justice, social pacts—while doing so little to promote equality. The reality of our system is that it protects those who have something and it is very tough on those who don't. It is time to rethink our economic and social model—not to destroy it but to renovate and improve it.

Is France too cheap when it comes to social spending? In fact we've never spent so much—social expenditures have risen from 20 percent to 33 percent of GDP since 1980—and the results have never been so unsatisfactory. We've got to break the vicious circle which consists of using the failure to get results as a reason to increase spending, over and over again. Unless we really rethink our approach, increasing spending and accumulating deficits and debt will not lead to better results. This is not an opinion based on some ideological position. The facts lead inevitably to this conclusion.

Is France too liberal economically? It has the highest rate of taxes and public expenditures as a share of GDP of all the large industrialized countries, the most restrictive labor laws, and one of the most developed systems of social insurance.

Is France a victim of globalization? Alas, our economic and social problems largely predate the beginning of globalization. These problems have been getting worse since 1981, which is now a generation ago. Our annual growth rate loses a half a point every ten years (2.5 percent per year on average during the 1980s, 2 percent per year during the 1990s, and 1.5 percent per year since 2000). French economic growth, consistently above the world average prior to 1990, has been below the world average since then. In 2004, world economic growth reached its highest level in twenty years (5.1 percent), and it was still at 4 percent in 2005, compared with 1.6 percent in France. The disturbing state of our foreign trade, long one of our strengths, is the result not of rising oil prices but of the poor placement of our products in international markets. Germany trades under the same foreign economic constraints as we, uses the same currency as we, and is even more dependent on energy imports than we, but had a record trade surplus in 2005. That was the year our deficit hit a historic high of €26.4 billion, compared with €8.3 billion in 2004.

Our difficulties are internal. More than 60 percent of our jobs are not even exposed to globalization. These sectors include retail, handicrafts, tourism, agriculture, environment, home maintenance, personal services, transport, energy, health care, and public services. None of this prevents us from making the State more effective, modernizing the civil service, reforming the education system, improving research efforts, or putting in place a policy for cities.

Between 1980 and 2004, France went from sixth to seventeenth place within the OECD (Organization for Economic Cooperation and Development) in terms of per capita GDP, that is, in terms of its living standards. The average French person thus got poorer compared with the inhabitants of other developed countries. Countries that were far behind us, including Ireland, Austria, the Netherlands, Belgium, the United Kingdom, and Finland, are now ahead.

I can already hear some out there claiming that GDP per capita is a purely economic criterion that does not really reflect a country's well-being. This is the view that will be spouted by the theoreticians of indifference, shouting from their isolated sheep farms in Larzac. It's also the opinion of the professional optimists of the government bureaucracies, blinded by the magnificence of the ornate old palaces in which they work. In their view, who cares about mediocre economic performance if it's balanced by a more humane social climate, high-quality public infrastructure, and the *art de vivre*?

What matters, actually, is quality of life. Let's take as a criterion the Human Development Index (HDI), which the U.N. calculates every year. This indicator combines and weighs three criteria: standard of living (GDP per capita), health and longevity (citizens' life expectancy at birth), and education (adult literacy rate and elementary school, secondary school, and university attendance). As it happens, these are all clearly social criteria. According to this indicator, France fell from eighth to sixteenth place in the world rankings between 1990 and 2003. The conclusion couldn't be clearer.

I'm not saying this to try to offend anyone or to stir up controversy. I'm saying it because we have to accept this reality and stop deluding ourselves. We're falling behind in the rankings of great nations. That's not particularly pleasant. But there's a reason: we're not trying as hard as others to adapt, modernize, and challenge ourselves. This is at least reassuring, since we know what we have to do to climb back up the ladder. There are no longer any excuses not to do it.

The truth is that for thirty years, France's ability to create wealth has been diminishing (economists would say that its potential growth rate is falling). It has allowed costly social problems to develop, and it no longer knows how to help anyone, because it has to help everyone.

SUCCESS AND INITIATIVE HAVE BEEN DISCOURAGED

How did we get to this point? We got here by putting off necessary reforms. We need to reform a government that costs too much while not always being effective, and we need to reform a social insurance system that wastes too much money. Between 10 percent and 15 percent of health insurance expenses are fraudulent. If we were to eliminate them we could get rid of the deficit in this area.

We also need to reform higher education and research to compete in an international environment in which knowledge and innovation are critical. We're really falling behind in this area. In an international ranking of universities done by a university in Shanghai, the top French entry came in at forty-sixth place. The École Polytechnique, the crown jewel of our higher education system, the school we're most proud of, is ranked between 203rd and 300th place (after the 100th place the schools are ranked by group). So there's our school, the one that opens our annual Fourteenth of July parade, modestly ranked alongside a hundred universities we've never even heard of!

France has been discouraging initiative and punishing success for the past twenty-five years. And the main consequence of preventing the most dynamic members of society from getting rich is to make everyone else poor. By trying to ensure equality for everyone you end up penalizing everyone. In his book *Les Français:*

Réflexions sur le destin d'un peuple (The French: Reflections on the Destiny of a People), former president Valéry Giscard d'Estaing reports his sincere amazement when François Mitterrand told him in 1984 that his objective was "to destroy the French bourgeoisie." Why would success and merit generate such hatred? The Socialists' dream is a society in which everyone is paid the minimum wage. My dream is a society in which someone who climbs the social ladder or wants to work is helped to find a job to fulfill his or her desire for social advancement.

The French problem with money actually goes even farther. What we're seeing is demonization and deification at the same time. For some, money is just corruption. It corrupts sports, politics, and business in general. It buys everything, corrupts everything, and destroys everything. François Mitterrand cleverly took advantage of this line of thinking when he talked about "evil money." For others, money means happiness. It makes everything possible, even effortless. Earning money ends up being an obsession. These extreme, but not necessarily contradictory, attitudes demonstrate France's discomfort with material success. Rather than serving as an example, success is often suspect. It's foreign, and ultimately illegitimate.

We've got to stop looking at failure as permanent, and we should encourage individual success, which lifts society as a whole. Money is nothing other than legitimate compensation for doing extra work or taking risks. It is a way to create wealth that leads to more growth and thus more jobs. The enduring French ideology about money and success leads only to impoverishment, leveling off, and egalitarianism. In other words, it leads to nothing either morally satisfying or effective. We're going to have to change in this area as well.

For twenty-five years, as our wealth has diminished, we have refused to make choices. By promising to help everyone, in the end we don't really help anyone. The employment subsidy (*prime pour l'emploi*—PPE) is for me the best example. Given out to no fewer than 8 million beneficiaries, it has lost its purpose. The money being sprinkled around doesn't make people any more satisfied.

We're trapped in the habit of creating a special status for everything. There are "minimum-wage earners," "single mothers," "unemployed whose benefits are running out," "youth," "handicapped persons," "teachers," "artisans," "civil servants," etc. Our social services are constantly busy figuring out who has the right to what and who goes into which file—when what they should be doing is taking an interest in individuals' specific problems. For example, one of the obstacles that prevent single mothers from going back to work is the lack of affordable child care. What are we waiting for, to help them resolve this problem individually or to guarantee them places in day care centers?

We have to move from virtual or theoretical justice to real and concrete justice. And to do that we have to learn to make choices and take responsibility for them. Let's agree to question some of our old habits. Not everyone has the right to the same "solidarity benefit," because some start with less than others. They need more public support. Choosing means giving more to those who, on life's starting line, start farther behind. I think that the employee of a company whose jobs are being exported has a right to more public assistance than a bureaucrat whose job is guaranteed for life. That's why I'm calling for change, for a French version of affirmative action in which certain regions and social groups would have the right to more support than others.

THE ABANDONED MIDDLE CLASS

Ever since the "Thirty Glorious Years"—the period from the mid-1940s to the mid-1970s when France experienced strong economic growth—we have gradually abandoned any social policy for the middle class. This is a mistake, because it's the middle class that ensures the prosperity and mobility of a society. Thus the middle class ought to be at the heart of any policy. Families are at the heart of the middle class. When the middle class prospers it helps society as a whole to advance. The current weakness of our society, and the cause of our general pessimism, comes from the fact that at the threshold of a new century the middle class has lost hope almost as much as the disadvantaged. Still rich enough to pay taxes and not poor enough to receive benefits, the French middle class faces numerous problems: job insecurity, falling purchasing power, and lack of affordable property and housing, higher education, and professional opportunities for its children. All this leads to a lack of confidence in the future.

Middle class people are asked to be more and more qualified. Yet they don't feel they're climbing up the social ladder. And in fact they have not advanced. When the middle class is stagnant, all of society becomes stuck. It becomes increasingly difficult to get better housing. Housing changes hands less frequently, preventing young people from moving in, and making it harder for those with the greatest needs to get cheap public housing. Those on the bottom can no longer hope to move up, and huge inequalities develop between the rich and all those whose situation fails to improve or even regresses.

I did not buy into the spirit of Jacques Chirac's 1995 presidential campaign, and not only because the candidate I supported, Édouard Balladur, did not get past the first round. It's not that I'm in favor of the "social fracture" or against help for the most disadvantaged; such a position would obviously make no sense. But to base a presidential project on the least dynamic part of society, that which is in most difficulty, is not the way to create the momentum necessary to mobilize society to move forward. The focus has to be on the middle class.

WORK IS NOT APPRECIATED

Finally, we made an enormous error in not placing enough value on work. Unfortunately, responsibility for this is widely shared.

I recently reread an article by Raymond Aron that appeared in *L'Express* in December 1982. It was called "The Power of Misconceptions." Written while François Mitterrand was trying all sorts of measures to share jobs, it explains in a few paragraphs the reasons this kind of policy is not the answer to unemployment. First, there's the immediate issue of paying for it: the public purse has to support whoever is not working. You can give your job to a young person, but then you have to share the salary. The next reason is that when you artificially reduce the overall amount of work done by society— whether by reducing the workweek or excluding from the workforce certain categories of workers such as senior citizens or single mothers—you necessarily also reduce the production, that is to say overall prosperity. Certain needs remain unsatisfied and some goods or services no longer find markets, because society's overall purchasing power is reduced. If the story is true, restaurant owners in Chamonix were absolutely right to have refused to serve former labor minister Martine Aubry at 9:30 one night, since keeping the restaurant open at this hour was no longer profitable because of the thirty-five-hour workweek that she sponsored! Unfortunately, twenty-five years after Mitterrand, people are still thinking the same way.

A number of countries—including France after the Algerian War—have experienced a massive population influx that can create jobs rather than unemployment, so long as it's accompanied by the necessary infrastructure investments (such as machines, offices, factories, and computers). The answer to the unemployment problem is work.

It's thus hard to exaggerate the damage done to France by the thirty-five-hour workweek. How can anyone think that you're going to create wealth and jobs by working less? After the mirage of the first few months, when people had to quickly hire new employees to keep open their hair salons or production lines, the effect of this reform on business competitiveness started to be felt and orders started to fall. Increasing social spending became necessary, salaries were stagnant, and buying power fell.

Cutting social charges is no longer enough to compensate for the gap between the productivity of unqualified workers and their pay, which must be compared with that of other workers and workers from emerging countries. Unemployment among unqualified workers, which is the hardest kind to fight, is getting worse. Finally, the cost of capital is falling relative to the cost of labor, and machines are replacing people whenever possible. On top of all the catastrophic effects of the thirty-five-hour workweek you have to add the disastrous effects on public services—for example, the hospitals that have been thrown into turmoil because of a reform that is as aberrant in theory as it is in practice.

I don't think the French have lost their desire to work. On the contrary, they are exasperated because welfare now pays better than work. Unlike being on welfare, working doesn't get you reduced prices, housing assistance, Christmas bonuses, health care coverage, breaks on housing or television taxes, or cheaper tickets for the movies or museums. On the contrary, work is penalized because the more you save for your children, the more inheritance taxes you pay. Working is no longer a guarantee of social advancement, or of opportunities to acquire property. Working doesn't even let you look calmly ahead to the prospect of retirement or possible health problems, because we are unable to say how we'll finance our social security system in the coming thirty years. The French people know this. It's not the development of a leisure society or some sort of modern Epicureanism that has broken France's dynamism and led the French gradually to lose interest in work and business. France has a working culture and farming culture. Its people know what work is. The French are not afraid of work. But the deliberate inversion of values between work and welfare has caused people to lose their bearings. When someone who works doesn't live any better than someone who doesn't work, why should the one working get up early in the morning?

I trace the irresponsible behavior of certain leading industrialists to this same inversion of values. Let me explain why. When company bosses pay themselves salaries, stock options, and retirement benefits at levels the equivalent of centuries of the minimum wage, it's because they have lost respect for their workers, themselves, and their companies. It's because our society has lost its reference points. When I call this behavior outrageous I'm actually understating it. You would never have seen this from a company president during the economy's glory years.

THE CHOICE: WORK LESS OR EARN MORE

The Socialist Party has a social program. It is based on solidarity and cohesion, and it promises to help the weakest members of society and restore hope to the middle class. But it's a lie. This program extends and worsens everything that has been failing for twenty-five years. It extends the thirty-five-hour workweek and retirement at age sixty when all our neighbors are doing the opposite. It continues to ignore world realities and the challenges of our time. It widens the gap between economic realities and social policy, which is the surest way to fail to provide both efficiency and justice. The Socialists propose to keep using the same methods, recipes, and ideas, those that have destroyed our economy and led our society toward doubt, disarray, and fear. This needs to stop. The Socialists want us to work less. I want us to have more buying power.

Nothing is more important than restoring work as a cardinal value. And to do that there is only one solution: proving that work pays. We must show that work brings greater remuneration and greater respect. Nothing is more discouraging than noticing that working more or taking more responsibility gets you nowhere. The benefits of working must in any case be greater than those of welfare. This is a matter of our country's progress and survival. Everyone must be certain that the path to social advancement is open to those who deserve it. I propose that we help anyone who wants to work provided that he or she demonstrates, through a minimum of effort, the desire to advance. The State cannot do for you what you're not willing to do for yourself. We need justice from above, not justice from below. In the first case we're encouraging prosperity; in the second we're leveling it off. I've made my choice, and it's the choice for social advancement. I believe in giving hope to all that by working hard they can live better than their parents and that their children can live better than they do. All should be able to look to the future calmly and happily.

CHAPTER 7

Remaking French Politics

THE CLEARSTREAM AFFAIR

It took me a long time to take the Clearstream affair seriously. Clearstream is a bank in Luxembourg that allegedly paid kickbacks to some French politicians and industrialists with money that came from a corrupt sale of French frigates to Taiwan. I had nothing to do with all this, so paid little attention at first. But the more I got to know about Clearstream—and especially once my name was outrageously dragged into the mud—the more I realized how serious it was. Personally, it was a lesson for me in how costly it can be to serve your country by running for national office. More generally, it's an example of the sort of scandal that France really must put an end to, forever.

I do not intend here to make any revelations or take any sort of political revenge. Politics—real politics—has nothing to do with this affair. If political leaders took part in it, they would not be politicians but simply liars.

I'm not naive and I know how tough public life can be. I know that all sorts of dirty tricks exist. I can even accept that my adversaries, or even my "friends," take advantage of my troubles. This has after all been done with no sense of decency or limits with my private life. I could do without this, but that's the way it is, and there's no point in complaining.

But in the case of Clearstream, we're talking about something else. A line was crossed, and we descended to a new and unacceptable level of villainy. It's no longer just a question of taking advantage of someone's difficulties, but creating them entirely from scratch with the sole aim of destroying somebody.

I know the history of the French Republic, and I mean all the gory details. I know the ugly part of its history and all the demeaning things that have been done. I know there are plenty of examples of such things. But that doesn't justify this affair. It's no reason to minimize what happened, to close our eyes, or to tolerate this sort of behavior. I'm using such a strong word—"tolerate"—on purpose. People have been tolerating this kind of attitude for too long. Everyone knows what's going on. But people just don't talk about it. They don't want others to think, "They're all corrupt." They don't want others saying, "There is no smoke without fire." They're afraid of breaking with a circle that too often demonstrates the worst sort of solidarity. They're afraid of being considered snitches. They're afraid of scandal and its political consequences. They're afraid people will conflate all sorts of unrelated issues.

I want to say that I'm not afraid. These plotters and schemers in their smoke-filled rooms have been poisoning politics for too long. They revel in plots. They live off of the lowest of human feelings—jealousy, hatred, and greed. By "they" I mean all those who are ready to work for whoever happens to have power at any given moment. Their role is to do the dirty work or even to anticipate the wishes of their masters. The cleverest of them thus manage to carry out orders without even having to be asked. That way the person behind the operation doesn't have to feel guilty.

I learned about the Clearstream affair through a telephone call from the director of the Internal Security Agency (DST) to my chief of staff. He was hardly going out on a limb: this was two days before the weekly magazine *Le Point* was going to put the story on its cover in July 2004. I admit that I didn't give it much thought. To me it seemed crazy that my name would be listed on alleged bank accounts at a bank in Luxembourg that I had never even heard of. I decided to treat the story with utter disdain, as I was certain that it wouldn't go anywhere. It didn't take long for me to be proved entirely wrong.

Things became more troubling when the investigating magistrate Renaud Van Ruymbeke got involved. I was in Moscow when two newspapers, *Le Journal du Dimanche* and *Le Parisien*, published a story saying that I had been the subject of two international letters of request *(commissions rogatoires)* sent by a magistrate who was known worldwide for his fight against corruption. Now it was becoming much more serious than I had realized.

A few months later, reading the letters of request, I went cold. The judge was not operating with kid gloves. I was suspected of having received kickbacks from the 1991 sale of frigates to Taiwan. This made no sense, since I had become minister in 1993, two years after the contract was signed. The judge wanted to know if I had used my alleged Clearstream accounts to launder the profits from this corrupt dealing. It took nineteen months of investigation to prove that all this was false. It's a good thing Van Ruymbeke is known for his competence, because I would hate to see how this would have played out if this wasn't the case. He must also be very busy, because in nineteen months he never took the time to inform me about his investigation. He certainly had no legal obligation to do so. But this affair was so unusual, and the evidence so thin, that the appropriate thing to do would have been to question me before starting this sort of legal procedure. Nor did he bother to let me know that the two letters of request had come back and that both turned up nothing. I later learned that the judge had the time to meet with Jean-Louis Gergorin—the author of an anonymous letter falsely implicating me in Clearstream. I concluded from this that Van Ruymbeke should think more carefully about his contacts and about how to make good use of his time as a magistrate.

Once I learned from the press what Van Ruymbeke was up to, I decided to follow this affair more seriously. I learned that the Internal Security Agency, the DST, also knew the details of the case, even though the justice system had never asked it to investigate. "Someone" asked it to make an inquiry, it knew that I was innocent, and it knew the source of this slander, of which, I might add, I was not the only target. I thus requested that the DST inform law enforcement officials according to Article 40 of the penal procedure code. This article requires any official who, while carrying out official functions, learns of a possible crime to inform the investigator. The DST refused to do this, despite my urgent request. I still don't understand why. But this only reinforced my determination to know what was behind this affair. And I would add that the requirement in Article 40 of the penal procedure code applies to all officials, including ministers. If the list of alleged Clearstream account holders had not been all or partly fake, it would have been a serious matter implicating senior political leaders or public figures and threatening national interests. The normal reaction of a minister who learned about such a matter would be not to tell a lone adviser but to inform the legal authorities so that they could put appropriate means behind an investigation.

Finally, in January 2006, I learned from an article in *Le Figaro* that the letters of request had come back, definitively clearing me. It was only at this point that I was able to lodge a civil suit. This was necessary in order to see the file on the case. How can you demonstrate the reality of a slander unless you have the means to explore what was behind the slander?

This was not an easy decision for me to make. Many observers saw it as a political move. They thought I was using this as part of my alleged "rivalry" with Dominique de Villepin. Anyone making that assumption misunderstands me. I have never used such methods and I never will. They disgust me. The real story is simpler. I want to know the truth. I want to understand. I want to get to the bottom of the intrigues and compromises behind this affair. I want to confound those who sought to impugn my integrity. I also want it to be clear that from now on, in the French Republic, such behavior will not go unpunished.

I have never met Jean-Louis Gergorin. Nor do I know his collaborator Imad Lahoud, who hacked into the Clearstream accounts, or Philippe Rondot, the former intelligence agent who investigated me at the request of Dominique de Villepin, an old friend of Gergorin. I learned about these colorful personalities from the press like everybody else. The worst in my mind seems to be Gergorin. Here's a man who doesn't know me yet who triggers an investigation of me, accuses me before a magistrate, and talks freely about a trip to India that I never took. The most intolerable part is that he carries on with his accusations and dirty tricks behind a facade of intellectual good sense. This impresses only those who are dim enough to give credibility to individuals who don't have any. Imad Lahoud is in my mind summed up rather well by his legal past: investigated for having stolen from his father-in-law! As for this Mr. Rondot, the way he keeps records hardly makes you want to know him or become friendly with him. When I think that his job is to work in secret, it makes me quiver. With him, you wonder about the usefulness of the secret services. Here was this famous group at work. I really think these rogues didn't even know how serious this was. You have got to wonder what interest someone like Dominique de Villepin found in seeing them.

And then there's the question: Were they working for someone? And if so, for whom? These questions weighed on me for a long time. They no longer do, now that I know that this affair will be fully investigated. It's our justice system that will reveal the truth. And it's then, and only then, that I will pursue the political consequences.

RESPONSIBILITY AND BALANCE IN THE WORKING OF THE STATE

The French Republic needs to recover a sense of discipline, a sense of responsibility, and a desire to build the future that it seems to have lost.

The left has its sacred cows, and so does the right. Among them is our famous Constitution of 1958, conceived of by General de Gaulle and written by Michel Debré. In every political science institute in France, the scholars repeat over and over again that the political system of the Fifth Republic is the best we've ever had and that it has given us the stability we so badly needed. Much of this is true, but let's be clear: we're only superficially stable. As I noted before, our prime ministers change much more frequently than those of other countries, and no parliamentary majority has been reelected in France since 1981.

The Constitution of the Fifth Republic was excellent in 1958. It enabled General de Gaulle to undertake numerous reforms, the most important of which was decolonization. Today, it is showing its weaknesses, perhaps even malfunctioning. The executive branch has too much power, a reality that is somewhat hidden because this power is shared by two different people. The founding principle of the Gaullist Republic—the notion of responsibility—was abandoned in 1986 when François Mitterrand, clearly rejected by the electorate, remained in power and accepted "cohabitation" with a government of the right. The 1958 Constitution, revised in 1962 to allow for direct election of the president, concentrates so much power in the hands of one man. To me, such concentration of power is justified only by political and moral responsibility. The French Parliament does not really play the role of counterweight, and it is exhausting itself daily with prolific, but hardly effective, legislative activity.

The guardians of the temple continue to insist that the problem is not the Constitution but the way we have been putting it into practice for years. Of course, individuals have played their part. And I am the first to admit that the main reasons for France's current disarray are its immobility, its lack of debate about ideas, and the absence of reforms, not the institutions themselves. Besides, the institutional changes I'm calling for are both textual and behavioral. But we can't wait for the arrival of some providential political class—which has never existed and will never exist. And we can't refuse to touch one word of the work of the founding father. If we do that we'll remain handicapped by a political system that is now in need of modernization. In my view, this modernization must focus on two priorities: responsibility and balance.

THE PRESIDENT AND THE PRIME MINISTER

The division of the executive branch between the president of the Republic and the prime minister makes the system more opaque and therefore dilutes responsibility. No one believes that the president doesn't really govern. No one believes that the prime ministers of Presidents de Gaulle, Giscard d'Estaing, or Mitterrand had much autonomy. And everyone knows that it was this lack of autonomy that led Jacques Chirac to resign as prime minister in 1976. Similarly, the strained relations between François Mitterrand and Prime Minister Michel Rocard were familiar to all. The reduction of the presidential term from seven to five years, in 2000, made the president even more powerful relative to the prime minister for two reasons. First, with a shorter mandate, the president must get even closer to the concrete, daily problems of the French. Second, holding the presidential and parliamentary elections almost simultaneously means that the fate of the president of the Republic is intimately tied to that of the government. The president therefore cannot avoid providing leadership to the government. Besides, in practice this was already the case: not a single word of a government's declaration of general policy is put forward without submitting it first to the president of the Republic. Not a single important cabinet meeting takes place in the prime minister's office without an adviser from the Elysée being present. Executive power lies in the hands of the president of the Republic because of the legitimacy that direct universal suffrage gives him.

This situation must be accepted as it is, in the clear view of the French people. The French must know who decides—and when, how, why, and in what circumstances. That is why I think that the prime minister's role should be recognized as one of coordination.

A number of strategic departments currently under the prime minister should be placed under the Elysée. Decisions should be made transparently at Elysée meetings announced in advance, not secretly or behind the scenes. These decisions must also be made more collectively. The presidential cabinet must no longer be a registry office for decisions negotiated by advisers; rather, it must be a place of discussion and debate, where firm decisions are made, and where the government expresses a collective will. Within the government, disagreement should be seen not as a big deal, but rather as an opportunity to make decisions that are better, more effective, more comprehensive, more balanced, and easier to understand. Finally, the president of the Republic must be able to come to Parliament himself to explain his policy rather than send messages to be read by a third party to members of Parliament standing at attention.

I am not talking about concentrating all power in the hands of the president or making him even more powerful by eliminating the prime minister's counterweight role. I am talking about recognizing today's reality and having the person who decides also be the one to take responsibility for his or her decisions. And if you need an institution to counterbalance the president of the Republic, this can only be the Parliament. The prime minister clearly does not have the means to do so without going beyond his role and preventing the president from leading the country.

THE *DOMAINE RÉSERVÉ* IN QUESTION

In exchange, certain powers of the presidency should be reduced, better supervised, or even eliminated. This sounds like a step in the opposite direction, but for me it's not, because the goal of all these measures is a presidency that is less monarchical, more transparent, more modern, and more democratic.

The existence of a *domaine réservé* of the presidency, in which neither the prime minister nor the Parliament nor the political parties have any say, is in my view incomprehensible and undemocratic. This is mainly the case for foreign affairs and European affairs. I have always thought, for example, that the European Union made a mistake by allowing the Central and East European countries to join before reforming EU institutions. No one can doubt my affection for these countries and their cultures. Enlargement was necessary. But the reform of EU institutions was also necessary. I think it's unfortunate that the issue of the pace of enlargement was not debated politically, particularly in Parliament. It is after all a really important issue, for France and the French. Now we have to make up for lost time lest we really weigh down the European Union. The only answer is to negotiate a shorter treaty, limited to the clauses that are essential to the functioning of the institutions. Nobody objected to these clauses during the referendum campaign. This new treaty would be ratified by the Parliament.

The prospect of Turkey's entering the European Union is equally nonsensical in my view. I understand the strategic hopes that underpin this idea. They could be achieved with the conclusion of a strategic partnership with Turkey. But 98 percent of Turkish territory is not in Europe. In twenty years, Turkey would be the most populous country in the EU, and the Turkish population is majority Muslim. Thus you have to admit that EU membership for Turkey would be such a shock that the Union would become a completely different organization from what it is now. It would be even more different from the founders' project of an integrated political Europe. If Turkey entered the EU, I also wonder on what basis we could exclude Israel, which has such close ties to France and Europe—or even Tunisia, Algeria, or Morocco, which were French a half century ago. Europe would have no more limits. It would become a subregion of the United Nations. The notion of a politically unified Europe would be finished. And yet, on this critical question, no debate took place in Parliament until October 2004, and even then it wasn't put to a vote.

Last October, less than six months after the French and Dutch rejections of the constitutional treaty, the European Union opened accession negotiations with Turkey. No head of state or government opposed this even though the role of the Turkish question in the results of the referenda is widely known and accepted. Europe seems incapable of escaping from the promise it made in 1963 in an entirely different geopolitical context. But the more time that passes, the harder it will be to tell the Turks that they cannot enter the EU. And yet this is a likely scenario, since the French Constitution now requires that Turkish accession be put to a referendum in France. I also feel strongly that we should have the courage to insist that Turkey fulfill its historical duty toward Armenia.

THE NEED FOR PUBLIC DEBATE ABOUT DEFENSE

We should have more public and political debate about defense policy, spending, goals, and results—particularly in Parliament. This is clearly compatible with the need for secrecy in these matters. I had a big debate with the president of the Republic on this in July 2004, and there have been so many erroneous reports about what I said that I'd like to clear it up here. I'm not revealing classified deliberations from the Defense Council, but I want to clarify exactly what I said.

I have never disputed the need to invest resources in France's defense that are equal to our ambitions. We spend 1.8 percent of GDP on defense compared with 2.2 percent in Great Britain, whose GDP is in addition bigger than ours. Unfortunately, this investment is now even more necessary because the Jospin government failed to invest adequately, preferring to treat itself to the thirty-five-hour workweek and to increase public spending rather than prepare for the future. I simply observed, first as a citizen and then as finance minister, that practically all defense investment programs were falling behind in terms of costs and schedules. This situation is in nobody's interest. It's obviously bad for taxpayers. But it's even worse for the military, which complains about delayed upgrading of equipment and which must spend more and more on maintenance of decrepit equipment.

As minister of finance, I was responsible for carrying out the 2004 finance law and for preparing the 2005 law. In a really tight budgetary context, and seeing how the 2003–2008 military program law got out of control, I wanted to understand what the problem was. This is doubtless a weakness of mine—I hate not understanding things. That's when I discovered that the Finance Ministry bureaucracy, in particular in the budget office, has no specific information or expertise about the defense budget. Under the cover of the *domaine réservé,* the size, management, and assessment of the defense budget are all pretty much a black box. It is not the subject of any critical evaluation or debate. The worst thing in all this—and I'm quite conscious of what I'm saying here—is that the main victim of this lack of transparency is the president of the Republic himself. How can he make decisions if he's not exposed to any serious discussion of alternative options to enlighten him about the military's big investment proposals?

I think defense questions should be debated within the government and the Parliament. These are important questions for the future of our country, and the budgetary stakes are so high that we should talk about them and give ourselves the means to manage our military investments effectively. This is also the best way to ensure public support for our defense policy and for the decisions that are made.

THE POWER TO NOMINATE AND PARDON

I think the president of the Republic has too much nominating power. On this issue, too, there is so much hypocrisy. We're always happy to praise our politically independent and neutral civil service and to criticize the American "spoils" system, which lets politicians replace all senior officials every time the government changes. The reality of our nomination procedure is unfortunately not so glorious. On paper it's transparent, with openings published in the *Official Journal.* But in reality it's opaque, and the choices are made even before these announcements are published. The personal preferences of the person making the nomination often play a bigger role than the abilities of the person nominated. For the most important posts, such as the Constitutional Council, the Superior Audiovisual Council, other independent departments, and the presidencies of public companies, the Parliament should be involved in the nomination process. A parliamentary commission could organize public hearings for candidates and could ratify the executive's proposals in a vote. This would better guarantee the competence and impartiality of those who put themselves forward for top administrative jobs.

Finally, I think we have to put an end to the president of the Republic's power of amnesty and pardon. This power is problematic where the separation of powers is concerned. And it's morally offensive in a Republic where citizens are supposed to have rights and duties. Chirac's amnesty for a well-known former professional athlete, Guy Drut, hurt the Republic because it gave credit to the idea of privileges reserved for elites. Drut's personal qualities are not the issue. This is a man who incarnated France in a remarkable way. But this amnesty was a Pyrrhic victory for him. He deserves better. I therefore support abolishing this presidential power, which dates from a bygone era. I can't really think of any good reason to maintain it. These sorts of practices make French democracy seem foreign to many around the world.

This would result in a president who governs; who is not afraid of debate within his government; who submits major defense, foreign policy, and European policy issues to Parliament; and who would come and explain his policies to the country's representatives. And in my view such a president would have enough to do that he could change jobs after two terms. I think that the more you focus on staying in office the less you govern. Thus I think that any president of the Republic should be limited to two terms. That should be enough for a president who wants to act and acceptable for those who think that the Republic periodically needs new blood.

MAKING PARLIAMENT A REAL COUNTERWEIGHT

Unlike Britain, which remains the model of a parliamentary democracy, France has always found it difficult to balance the different branches of power. It either gives too much power to Parliament or gives too much to the executive. This may be because, deep down, France is not a very liberal country, in the political sense of the word. It doesn't have the same passion as Britain for seeking as much freedom and independence as possible for its citizens. The British expect the law to guarantee these freedoms. The French expect the law to solve society's problems.

Too many powers are now concentrated in the executive branch in France. I don't think that a better democratic balance can result from this unclear and continually changing division of labor between the president of the Republic and the prime minister. I do think our Parliament should be stronger in order to be a counterweight to the president.

To be sure, some political experts would say that the same majority is in charge of both the executive and the legislative branches. In other words, there's no point in reinforcing the role of the Parliament, since the Parliament and the president come from the same majority. Today, they would say, Montesquieu's separation of powers is to be found not in a better balance between the executive and legislative branches but in a balance between the majority and the opposition. In fact, one of the main characteristics of the British system is the official status of the opposition, which allows it to play a balancing role. The British opposition has the means at its disposal to organize a meaningful shadow cabinet. It also has its own legal rights, such as, for example, that of chairing the parliamentary oversight commission on public spending.

All of these are good points, but I would respond in three ways. First, executive power and legislative power are in the hands of the same majority in Britain, just as in France. Second, reinforcing the Parliament's rights necessarily means reinforcing the rights of the opposition, because there are always members of the opposition in Parliament. And finally, the Parliament, given the way it is selected, has the real virtue of representing the diversity of French society better than the executive can. The executive is essentially a single individual surrounded by a few ministers. There are sometimes more differences between two members of Parliament from the same party but from very different districts than there are between two members from different parties elected in sociologically similar districts. There is thus nothing like the Parliament for expressing what is felt by the population. It is uniquely placed to find compromises among conflicting interests. In modern democracies, open and complex as they are, collaboration and compromise are not signs of weakness but rather conditions for reform. The role of politics is increasingly to find ways to reconcile conflicting points of view and to satisfy diverging interests.

Every time I put forward a bill in Parliament, I always sought to involve members in its development and to accept their amendments. Sometimes I even worked on the drafting with them. I even accepted a number of amendments from the opposition. For example, at the time of the vote on the 2006 counterterrorism law, an amendment from the Socialist group in the National Assembly led to the creation of a parliamentary commission to monitor the intelligence agencies. This kind of openness makes it possible to pass laws that are more balanced, more operational, and ultimately better accepted by their erstwhile opponents. I've always been bored by artificial debates and by the sort of parliamentary jousting in which everything the right says is jeered by the left and vice versa. I also know that France would benefit from having a government that included people from the other side, united by the will to change things. This could be arranged through an agreement based on mutual trust for the length of a parliamentary term.

Our Parliament is currently weak. It has limited means. The support structures for French parliamentarians cannot even be compared to those of American members of Congress, who are surrounded by proper staffs. Our Parliament is deprived of information. I would like to understand why, for example, the decisions of the Council of State on government-sponsored bills and the impact studies prepared for public officials are not transmitted to Parliament. In case of a difference between the Parliament and the government, which sometimes happens and should not be a big deal, Article 49-3 of the Constitution places members of Parliament in a real dilemma: accept the executive's draft or bring down the government. These options are too extreme and they lead to the sort of debacle we saw with the First Job Contract (CPE) law in spring 2006. Finally, Parliament lives under the constant threat of dissolution, which limits its ability to oppose the government.

In any case, real power is political power. You could give members of Parliament all the additional means and procedures you want, and yet Parliament would become a counterweight to the president of the Republic and the government only when these executive representatives felt it necessary to negotiate.

To reinforce these powers, I think that Parliament should first of all be able to debate and pass resolutions on defense, foreign policy, and European policy issues. It should be able to pass resolutions in the areas of government policy that are not really covered by the law, in order to influence what the government does in these areas. Even though it was legally consistent with the current constitutional language, it was politically incomprehensible when the Constitutional Council denied the Parliament the possibility of setting guidelines for the government on policies regarding education, development aid, and immigrant integration. I think that we have got to abolish Article 49-3 of the Constitution and give Parliament more of its own resources. In particular, Parliament should be able to make better use of the Court of Auditors (Cour des Comptes) to monitor government activity, as is the case in a number of democracies, including Great Britain. Finally, it would be desirable for protest parties to be represented one way or another in the National Assembly or the Senate. A dose of proportional representation seems to me necessary. I think that this would make it possible to bring these parties closer to positions consistent with our core principles. You radicalize people not by including them but by excluding them.

In the same spirit, I think we have got to increase the resources of the opposition. As it is now, the party that loses the elections becomes the party with the most limited means. A more relaxed and balanced democracy would, on the contrary, demand that this party be helped to rebuild itself and to play the role of opposition. I therefore propose that the parliamentary opposition be given an official status. This would make it possible to recognize it as an indispensable actor in our institutions and to give it certain rights. These could include being involved in consultations during a time of crisis, meeting regularly with the president of the Republic, and having representatives take part in official visits. Moreover, the public resources given to political parties should be better spread out over time so that a party that loses an election does not lose too much of its capacity. Finally, sixty deputies or sixty senators should be able to demand the creation of an investigative commission, with each member of Parliament able to call for such a measure once per parliamentary term.

The role and powers of Parliament symbolize the proper functioning of a democracy. A powerful Parliament is a good sign of a lively democracy. We should not be afraid of giving extra powers to the elected representatives. There is no alternative to reinforcing the Parliament's power of initiative and oversight. In France it would be about time!

In exchange, the Parliament has to stop proliferating laws left and right. France, and in particular French companies, can't bear it anymore. The Parliament is not solely responsible for letting this problem develop. The government puts forward too many bills. And the Parliament tries to compensate for its systemic weaknesses by writing or amending laws. That's why so many laws are full of purely declaratory articles which have no business being there but which allow the Parliament to put its views on record.

We must restore discipline to the legislative process, with fewer, better-prepared, and better-drafted laws. There are a number of different ways to do this, the main one being to incorporate the principle of "legitimate confidence" into the Constitution. Strange as this might seem, this principle simply obliges the State to live up to its word. For example, if the State adopts legislation that is supposed to last for ten years, it cannot change that legislation before then. Similarly, I think that we should adopt the British practice of publishing "green papers" and "white papers." When the British government plans major reforms, it begins by drafting a green paper setting out the problem and different options. Then it sets up a working group whose mission it is to collect written and oral testimony from everyone who wants to weigh in on the subject. Finally, the government drafts a white paper in which it sets out the option that it plans to implement and more discussion takes place. At the end, civil society has been involved with the reform process and the laws are better prepared and much more relevant.

REORGANIZING GOVERNMENT

When it comes to the quality of laws, the prime minister has a critical role to play. It's up to the prime minister to end the tradition that every single minister must have a law named after him. It's up to the prime minister to oblige ministers to run their departments rather than hiding behind laws passed just to give the impression of activity. It's up to the prime minister to make sure bills are necessary and of the right quality. Finally, it's up to the prime minister to demand of ministers that implementation orders are ready before the laws are voted on in Parliament so that both can come into effect at the same time. I don't see any reason why the Parliament shouldn't have the power to step in and apply laws if the government has failed to publish the implementation orders after a given amount of time.

France as a country is only as dignified and credible as its government. Thus, our ministerial instability and the continual changing of ministerial responsibilities make us look ridiculous. The organization of the State is not a toy that presidents of the Republic can use as they see fit. We need to define the responsibilities of the important ministries in an organic law so that we avoid dubious creations such as the "minister for free time" that François Mitterrand thought up in the 1980s. In a country of 300 million inhabitants such as the United States, the secretary of the treasury's area of responsibility is the same whether the Democrats or the Republicans are in power.

The process of elaborating this organic law could be an opportunity to proceed with other reform measures that I think are indispensable. For example, we need to bring all issues related to immigration into the same ministry, rather than handling them in three different ministries as at present. Today, visas and asylum are handled by the Foreign Ministry, integration and ordinary immigration are handled by the Social Affairs Ministry, and illegal immigration is handled by the Interior Ministry. One could also mention the issue of nationality rights, which is handled by the Justice Ministry. Similarly, if we really want environmental policy to be more than a series of empty pledges, we have to give the Environment Ministry some real power. It should handle energy, transportation, and equipment policy as well as the issue of industrial risks. Reducing greenhouse gases, the number one environmental issue, depends essentially on our ability to cut carbon consumption, to build clean cars, and to use more human transportation (such as walking and bicycling) and public transportation (such as railways). Creating a big Environment Ministry such as this would also have the advantage of giving environmental policy its own decentralized department, staffed with dedicated government officials.

I think we have got to limit drastically the number of ministers. The right number seems to me to be around fifteen, which works perfectly well for most of our main partners. This would guarantee a government that would be more focused on real challenges. The ministers could be supported by younger state secretaries without portfolio. The latter would learn the job, help them out, and represent them in a range of ways.

Government-funded housing should in my view be provided only to executive branch leaders whose jobs require them to be available at all times. This obviously includes the president of the Republic and the prime minister. It also includes the interior, defense, foreign affairs, justice, transportation, and finance ministers. For all the others, free housing cannot be justified, and the French people are right to demand less spending and more discipline in managing the public purse.

NOT BURDENING FUTURE GENERATIONS WITH OUR FAILINGS

Exercising power responsibly means not playing around with the money of the French people today or that of future generations. As finance minister, I modified the organic finance law—our financial constitution—to require each government to indicate in advance what it would do if economic growth turns out to be higher than forecast. The government must indicate this at the time of the vote on the finance law. Lionel Jospin's practice of using the revenue from economic growth for current spending rather than for national debt relief was irresponsible in a country in such a difficult financial situation.

In 2003, Germany reformed its health insurance system and required each agency—the German system is much more decentralized than the French one—to balance its budget every year without borrowing, that is, without adding to the debt. In case of a deficit, contributions are raised to reestablish budgetary balance. This is a demanding rule. Its implementation is unpopular. But it prevents the government from penalizing future generations for its inability to finance the health insurance system. I would support France's experimenting with a similar rule. In case of a deficit in the health insurance system, the government would be obliged to adopt measures to eliminate it in the finance law the following year. This could be done through a rise in contributions, through an increase in the Generalized Social Contribution (CSG), or by gradually raising various exemptions or payments to try to control the evolution of health care spending. In case of a surplus, the opposite rule would apply and the payments made by those who participate in the system would be reduced. It is clearly possible to distinguish structural deficits, which must be eliminated, from temporary deficits that can arise periodically because of slower growth or an exceptional health care crisis. Thus it would make sense to spread out the financing over a number of years.

THE LEGAL RESPONSIBILITY OF THE PRESIDENT OF THE REPUBLIC

Finally, with the same concern for putting our institutional regime on the same level as that of other major democracies, we have to deal with the issue of the legal responsibility of the president of the Republic. The solution put forward by the Court of Appeals (Cour de Cassation), which suspends expiration of the statute of limitations so long as the president remains in power, seems appropriate to me, and it should be written into the Constitution. The commission that was directed by legal scholar Pierre Avril, who was nominated by Jacques Chirac, also proposed a new formulation of high treason. This new formulation would give the Parliament the ability to dismiss a president in the middle of his term for having very seriously failed to fulfill his duties. This is a delicate subject, because the definition of "serious failure" could become exclusively political. Moreover, I'm not sure that this French version of impeachment would really be effective. If it had been in place during François Mitterrand's time, would it have been invoked—for example, in the case of the illegal wiretapping he authorized in the 1980s? I'm not sure it would have been. Unfortunately—or in another way fortunately—the real protection of the people and of democracy from dirty tricks, intrigue, and malevolence is not to be found in formal procedures. Rather, it is to be found much more in discipline, accepted practices, and political leaders' sense of the State and respect for France. I sincerely hope that French democracy will turn its back on the sorts of behavior that dishonor it among nations and sap the confidence of its people. If the creation of an impeachment procedure could contribute to that, I would be for it.

ENSURING RESPECT FOR OUR LAW ENFORCEMENT SYSTEM

We must ensure respect for our law enforcement system. In my view there are three critical issues. The first concerns the responsibility of magistrates. This is not a problem in the vast majority of cases. Most magistrates do their job well. But if the errors of a few are never punished, no one can have confidence in the justice system. It should thus be possible to hold magistrates responsible for their errors or personal negligence, as is the case for other government officials. Thus a citizen who felt he or she had been the victim of a judicial error could take the issue to the Superior Council of the Magistracy (CSM). Obviously, a mechanism to avoid abusing the system would be necessary. The CSM should also be reformed so that magistrates themselves do not form a majority of its members. It is normal to want to shield the magistracy from political influence. But it would be better protected if it were supervised by an institution that reflected society as a whole rather than by one essentially run by the magistrates' union.

The second issue concerns the resources available to the justice system. The image of French justice is damaged, and the confidence of the people undermined, by how slow it is to act. It is also damaged by its overburdened offices, laws that are continually changing, overworked judges who have to do many things at the same time, and its dispersal among too many small tribunals. That's why we need to increase the system's resources and to reform its organization by regrouping tribunals, reducing the isolation of judges, and making jurisdictions more specialized. Specialized magistrates operate more effectively and more quickly. Just as we reorganized the geographic distribution of the police and gendarmerie, changing a structure that had been in place since 1941, we could redraw the judicial map as long as we work with all the interested parties. The goal would be a comprehensive reorganization of national public services. We would decide where to put tribunals, subprefectures, treasuries, and hospitals, rather than letting each region reorganize itself without taking account of what the others are doing, as so often in the past.

Finally, we have to find a way to preserve the government's ability to formulate law enforcement policy while ending its political influence over the individual matters it deals with. This unacceptable political meddling is worthy of a banana republic. This issue might seem simple, but it's actually very complex. The reason is that to have a law enforcement policy, which I think is indispensable, the government must not only be able to give general guidelines to the public prosecutor. It must also be able, when necessary, to give prosecutors instructions in individual cases. For example, during the crisis in the suburbs, it was unfortunate that the Chancellery did not step in to demand that the Seine-Saint-Denis prosecutor's office take tougher action against the youths who had been turned in. It is really hard to understand how, in the Seine-Saint-Denis department, the only person arrested during the riots was a police officer.

To enable the application of crime policy and at the same time guarantee that political leaders don't abuse their authority to protect their own, I suggest the creation of a position of a "national prosecutor." This senior magistrate, whose competence and stature would be unquestionable, would be named by the government, after public hearings before a parliamentary commission that could oppose his or her nomination by qualified majority. The national prosecutor would not be independent from the justice minister, but he or she would be in charge of looking after the day-to-day application of the government's penal policy, which is all the more important since the justice minister has a lot more than that to do. It would be the justice minister who would, if necessary, give the individual instructions to the prosecutors, only for purposes of the general interest or jurisprudential coherence. The existence of this filter against political influence would be a guarantee for citizens of good professional practices in the implementation of crime policy.

Many of the French were certainly shaken, as I was, by the distressed and distressing face of former Toulouse mayor Dominique Baudis when he appeared on television to explain—and to challenge the veracity of—some terrible rumors that were circulating about him. I immediately thought that he was innocent. But not everyone had the same reaction at the time. The worst possible injustice is probably being accused of things that are completely contrary to who you are. Justice is not a joke. Slander is serious because once it's out there it never fully goes away, especially these days when the most frivolous rumors circulate on the Internet, where they remain. Justice must be rapidly and professionally administered. Criminals should be afraid, while the innocent should have nothing to worry about.

PROUD OF FRANCE'S HISTORY

Like millions of French people, I was offended and hurt to hear the "Marseillaise" jeered at the start of the France-Spain World Cup soccer match. This may have bothered me even more than when the "Marseillaise" was jeered by the French themselves, at the Stade de France when France played Algeria. France's reputation abroad is not good these days. People don't like France's sense of self-importance, given the country's serious internal difficulties.

Fair or not, there have been so many complaints about our "arrogance" that it has become a handicap as the charge has gained in credibility. France is not liked enough because it's not likable enough. In the world today, all nations deserve the same degree of consideration. The smaller the country, the more it wants to be respected. We have not taken adequate account of this reality, which requires us to understand that others can have the same demands as we. We would do well to know how to make ourselves accessible to others. When you're confident about your strength and the appeal of your culture, you don't need to be arrogant, self-satisfied, or pretentious.

France needs to learn to distinguish between pride and arrogance. To say that our country has problems, that it is threatened with decline if it doesn't make the necessary efforts, does not in any way diminish its honor or the legitimate pride that France inspires among the French. On the other hand, elevating everything we do to the status of a model and preaching to the entire world is pure arrogance. Actually these two approaches are complete opposites. It is precisely because we refuse to address our problems head-on that our flame is diminishing in the eyes of others. It's why other nations' respect for France is declining.

For the past few years, certain opinion leaders and pressure groups have been trying to make the French doubt their past prestige. They're taking advantage of our problems. They make Hitler into an heir of Napoleon. They turn slavery into a symbol of France, forgetting that France was not the only country to use this barbaric practice and that France produced many men and women who fought it. They reduce colonization to a criminal enterprise, even though this period of our history is much more complicated than that and took place in a context completely different from that of today. They focus on the mutineers of the First World War without thinking about the millions of French who fought in the trenches for honor and freedom. They pit the French against one another by associating the right with the anti-Semitic archconservatives of the Dreyfus affair and the left with France's democratic roots and the Resistance.

This line of argument is destructive. First, it's not true. France has made mistakes. Its history has dark areas. But France has fortunately recognized these errors, including the slave trade. Most importantly, France has always done the right thing in the end. Today, no French person, on the right or on the left, contests the fact that the country's honor was in London, not at Vichy, and that it was on the side of Dreyfus, not his opponents. As Max Gallo has written in *Fier d'être français* (*Proud of Being French*), France has not produced a Hitler, a Stalin, or a Pol Pot. It has not produced the equivalents of concentration camps or the gulag or wiped any cities off the map with nuclear weapons. Its natural inclination pushed it more often toward defending freedom and human rights than toward less honorable choices. This is in fact why the French nation has been respected and admired around the world. The French can be proud of their history.

An even greater problem with this denigration is that it risks undermining the foundation of the French nation. In this country with many different faces, where northerners are much different from Marseillais and Bretons different from Strasbourgers, national unity is made of the love of our history, culture, and language. Since the end of the nineteenth century, this unity has enabled entire generations of immigrants to become French without in any way having to renounce their original culture. Instead they've added to the melting pot. Being French is not a matter of where you were born. It is a matter of recognizing yourself in the culture and history of this country of incomparable destiny. It is precisely for that reason that those who do not love France do not have to stay. If everyone were to focus on the suffering that France inflicted on his ancestors, without getting over it once the history has been clarified and the regrets have been expressed, then the Cévenols would have as many reasons as the Vendéens and the Martiniquais to secede. After people get through explaining to the French all the reasons there are for not loving France, what will be left of our nation's spirit? What is important is to build our future together, not to tear ourselves apart over the past.

I do not like the expression "people of the left" that is used so often by the adversaries of my political family. There is not a "people of the left" or a "people of the right." There is the French people. Its strength and unity come from its capacity to unite people as diverse as Clemenceau, General de Gaulle, Jean Jaurès, and Léon Blum within the same heritage.

France must be proud of its past. I insisted on saying all these things in my speech on France in Nîmes on May 9, 2006, because we have to put a stop to the unjust attacks, false accusations, and systematic denigration.

And France must give itself the means to be proud of its present. It must reignite economic growth and full employment. It must give hope to its people. It must rebuild solid institutions founded on responsibility and a sense of the State. It must deepen its democracy through the development of a culture of open debate, compromise, and a balance of power. It must exercise power in an ethical way. It needs to do all these things so that its desire to lead the world down an original path is no longer seen as arrogance. This French desire should instead be seen as the gift of a generous, open, committed country that wants to see the world evolve toward peace among nations and well-being for all.

CHAPTER 8

Our Complex Society

Many French people have lost hope that change in France is possible. They think that reform is impossible or that it can be done only through violent confrontation. This is not my view. I don't share the vision of such an ossified France and such conservative French people.

The French are not afraid of change. They're yearning for it. It's politics that has gradually become sclerotic, predictable, and rigid over the past few years, not society. On the contrary, society has been profoundly transformed. People look at social issues much differently today from in the past.

Today, for example, French people recognize and accept the sincerity of homosexual love. They support Civil Solidarity Pacts (PACS) and reject discrimination. Sexual orientation is not a choice but an identity; I really believe that. Inequalities based on this orientation would be offensive. I don't accept that. This does not, however, eliminate in any way my reservations about marriage and adoption for homosexual couples.

It's not the job of the French people to change to accept reform. They won't do it. It's up to the political leaders in our Republic to rethink their methods.

SOCIETY IS AHEAD OF THE POLITICAL CLASS

My conviction, or perhaps my gamble, is that the French are much more lucid about and receptive to the world that surrounds them than is often said. They understand that the world is changing rapidly. Contrary to what they're accused of in order to explain away official inaction, they do not expect everything from the State, far from it. They know that the transformation of the world requires them to be able to react and adapt. They travel more and more, are familiar with new technologies, and get excited by the millions about a soccer World Cup that encapsulates the positive side of globalization. The Erasmus program, which enables French students to spend part of their studies in other European countries, has attracted hundreds of thousands of French youth. One sort of proof of this popularity is the success of two films by Cédric Klapisch, *L'Auberge Español* and *Les Poupées Russes*. Watching these films, and without thereby underestimating the difficulties of the world today, I tell myself that our children are lucky to live in this peaceful and democratic Europe running from Brittany to the Urals. A number of talented young French people are coming up with great innovations, though unfortunately too often abroad. And most of our companies are opening up to the world and have no problem with economic globalization.

On the other hand there is a growing gap between this open, active, and modern France, and a public sphere that seems inert. The State is not being modernized and is increasingly offering only two overly simplified policy choices that are almost caricatures.

The first is to deplore globalization and suggest that we can reverse it. The question, though, is not whether globalization is good or bad; globalization is here. And the truth is that it brings as many new problems as benefits. Among the benefits are new ideas, cultural exchanges, scientific progress, the extension of democracy, and a spectacular reduction in the price of consumer goods, especially high-technology goods.

The second approach, just as mistaken, is pretending that we have time and that it is possible to put off answers, and that we can do the opposite of what our competitors are doing without hurting ourselves. This approach actually feeds the anxiety of the French. The avoidance of honest self-examination has ended up making the French think that the reality is being hidden from them. This reality must be even worse than they fear. A discourse meant to reassure actually ends up creating and maintaining fear.

THE INEVITABLE FAILURE OF THE FIRST JOB CONTRACT

The crisis over the First Job Contract (CPE) is in no way the proof—wanted by some and feared by others—that France is unreformable. In fact, the failure of this reform was inevitable.

It was inevitable not only because of the way the government proceeded, failing to consult others adequately and resorting to Article 49-3 of the Constitution to obtain the agreement of the National Assembly. In fact, the use of this procedure deprived the government of a real debate in Parliament that doubtless would have led it to realize its mistake sooner. This only confirms why it would be advantageous to eliminate this article. But the problem with the CPE was even more serious than that. It was a substantive issue. No young person can accept being fired after two years for the sole reason that he or she is less than twenty-six years old. I was sure that the CPE would be seen to be unjust for the simple reason that it was. After a few weeks, the CPE was rejected by the young and resisted by companies that didn't see its purpose and didn't believe in its viability. Those it was meant to serve didn't want it and those who were resisting it were becoming more committed to the fight every day. The French Confederation of Christian Workers (CFTC) took the same line as the General Confederation of Labor (CGT), and business leaders, worried about the social consequences, were asking us to give up on the law.

I regretted that we didn't do so sooner, since it was obvious that it was never going to be accepted. The CPE was not the reform that our labor market needed, even if it did make that market a bit more flexible. Nor would the CPE symbolize the success of Jacques Chirac's second term, like the reform of the retirement system or the planned reform in the area of security. On the other hand, it is clear that the CPE risked making a mockery of our ideas about making labor laws more flexible while preserving job security. Instead it increased the segmentation of the labor market and rejected social dialogue. The desire for equality is so strong in France that this system created great risks for us. You don't recover from the criticism that you are not fair. For all these reasons, I believed that the cost of supporting the CPE was much higher for the majority than the cost of dropping it.

I took the risk of standing up to the most determined part of our electorate because I was convinced that the right must not repeat the errors of the past. One of these dangers was appearing to confirm the stereotype of a right that equated job flexibility with job insecurity. By promoting the CPE, the majority was giving the left an opportunity to recover momentum that it had not had for four years. In this case, the issue was much less a lack of courage than it was flawed judgment. To force through a marginal or even useless reform would have been a good way to hand the 2007 elections to the left. We stopped the damage just in time.

The fact that the republican right has finally gotten over its hang-ups about being on the right does not mean that it should spoil everything by falling into the traps of caricature. The right must defend justice, equity, and balance with the same determination as the left. Moreover, it must convince people that the change it proposes protects our ideals and builds a more just society whereas standing still maintains injustices. My profound belief is that progress is now on the right while conservatism is on the left. But if we don't convince people of this we will not be able to win over a majority of the French. The issue is not just standing for generosity, solidarity, and fraternity but also, and especially, a certain conception of humanity, a certain ethic. This is what is at stake: going beyond the limits of our political family to create the conditions for a popular movement that will modernize France.

GETTING RID OF DOUBLE PUNISHMENT

I am proud to be the politician on the right who abolished double punishment in our legal system. The left dreamed of doing so but didn't know how to get it done. We got it done, demonstrating French society's openness of mind and the right's capacity to respond to the public's desire for both firmness and justice. Accomplishing this reform was an important step in my political life. I changed my mind on the issue. Some of my close friends were surprised. Political will made it possible to create room to maneuver that people didn't think existed. I must admit that I learned a lot from spending time with the adversaries of double punishment.

Double punishment refers to the law that allowed the judiciary or the government to send foreign criminals back to their homelands after they had done their prison terms—and it could do this regardless of the individual's personal and family links in France. Elimination of double punishment was one of François Mitterrand's "101 Proposals"—in 1981! Lionel Jospin's position, twenty years later, was also that it should be eliminated but that the French people weren't ready for it. At best, this was a weak conception of the role of elected representatives in a democracy, and at worst it was political cowardice. For the left, the strategy was simple: the more they talked about something the less they did anything about it.

I must in all honesty admit that I didn't plan on implementing this reform either. Like many French people, I thought that individuals lucky enough to be welcomed in France are in a way doubly guilty if they break the law. I therefore supported sending foreign criminals back to their homelands.

It was the case of Chérif Bouchelaleg that led me to think again about double punishment, to learn more about it, and then,

finally, to change my mind. Chérif Bouchelaleg had been convicted
in France for a number of different offenses. When he got out of
prison, he was supposed to be sent back to his homeland as part
of a sentence that included his expulsion from France. As is often
the case, the local and national press reinforced the pressure that
his family—a French wife and six French children—was putting
on the authorities to allow him to remain in the country. The official memo that I got on the matter was in complete contradiction with the information in the press. The official memo claimed
that Bouchelaleg no longer had any contact with his family, while
the press was saying that the family often visited him in prison. I
wanted to have a clear conscience about it so I personally called one
of the journalists who was following the issue.

After the journalist got over the surprise of my call—he must
have thought I was calling to criticize him—we had a long conversation about the case of Chérif Bouchelaleg and more generally
about the issue of double punishment. I came to understand that
the dry official reports didn't give a very clear picture of what this
measure meant. Double punishment was inhuman. It consisted of
sending back to their countries of origin people who were certainly
foreigners on paper, but who might have lived in France since they
were very young or may even have been born in France. These were
people who had no links to their countries of origin, and who most
often had French families. It was the family that was punished and
torn apart because the father, whatever his mistakes, was sent thousands of kilometers away. It was the spouse who was condemned
to live without the person she loved, and to bring her children up
alone. The children were the ones who grew up without a father.
The State itself was creating single-parent homes. I was minister of
the interior to punish criminals, and Chérif Bouchelaleg had done
his prison term. I did not become a minister to punish children.
And I also gave some thought to what these children would, for
their entire lives, think about the country that had separated them
from their father.

After doing some more research I also came to understand

that as a practical matter double punishment could not be implemented. The situation of the families was such that most of the people involved opted to remain in France illegally. Upon learning that, I resolved to get rid of double punishment. My conviction was strengthened by the fact that I had really investigated it personally, carefully, and over a long period. I didn't underestimate the obstacles to change, but I was sure that the cause was worth it.

The first obstacle was the very negative reaction of the justice minister and the foreign affairs minister. Their reservations were hard to understand. Yet they led to a number of problems with the handling of individual cases, since visa issues, residency permits, and judicial decisions are closely interrelated. The announcement that the legislation was going to be changed gave great hope to the families concerned, and several hundred cases had to be reexamined even before the law was changed. Finally, and most important, I had to convince the UMP and the Parliament. The size of the challenge became immediately clear just days after I announced my plans when the left decided to make things difficult for me by putting forward a bill ending double punishment. Even the Socialists didn't really intend to have a vote on this hastily drafted bill they had prepared without consulting anybody. But during the debate, while some deputies from the majority supported the abolition of double punishment, the majority of UMP parliamentarians expressed strong opposition. I was neither offended by nor worried about this stance. A few weeks beforehand that had been my position as well. Instead, I set out to convince them and public opinion that double punishment should be abolished.

To do this, I turned to the methods I believe in, methods that make democratic life noble and purposeful: listening, dialogue, debate, and appealing to public opinion through the media and an information campaign. I was able to rely on four exceptional people, each of whom was in his own way committed to ending double punishment. These were Jacques Stewart, the president of the immigrants' rights association Cimade; Jean Costil, a Protestant pastor from the Lyon region who had for nearly thirty years been working on issues such as immigrants' rights, double punishment, integration of North Africans, and asylum; Bernard Bolze, then leader of the national campaign against double punishment ("One Punishment, Period"); and Bertrand Tavernier, the well-known filmmaker and director of the wonderful movie *History of Broken Lives: The Lyon Double Punishments,* which tells the story of the fifty-one-day hunger strike of a dozen foreigners in Lyon in 1998. This film describes better than I ever could the consequences of double punishment for couples, children, and families, as well as the incoherence of our legislation, and the discomfort of a State that is conscious of the problems but incapable of solving them. I was really disappointed that Jean-Louis Debré, as president of the National Assembly, did not allow this film to be shown to the members of Parliament.

During my four years in government, I met a number of exceptional people, and it's frustrating to me that I can't name and describe all these deserving people here. But I am naming the four people mentioned above because they are an example of the best sort of marriage between political activism and intellectual honesty. I have no idea about the political opinions of any of them. I have a feeling that some if not all of their political colors are different from mine. But to be honest, it makes no difference. What matters is that they all had the courage to put their trust in me and to engage in dialogue, whereas so many others decided to stay away for the sorts of ideological reasons that are among the causes of France's inertia. When the working group that I named announced its conclusions, Bernard Bolze was honest enough to defend them and to say that they constituted unquestionable progress. I know that he wasn't always in an easy position. Indeed, certain interest groups tried to grab on to one or two hesitations of what was only a working group to challenge my good faith. When I ultimately put forward my reform plan, which went beyond the proposals of the working group, he didn't turn up his nose on the grounds that it was a minister of the right who was advancing the cause that he believed in. Later, when my colleagues asked Bertrand Tavernier to provide testimony for a short film that was being prepared for the November 2004 UMP Congress, during which I was to be elected president of the party, he accepted immediately. Without qualification, what he said was that I had made a commitment to abolish double punishment and that I upheld this commitment. I know a number of other people I have helped while doing my job who—had they been approached for the same sort of favor—would not have been as honest.

Thanks to the courage of people like this, the reform was ultimately approved unanimously in Parliament. It caused no concern among the French. Curiously, the president of the Republic never mentioned the abolition of double punishment among his accomplishments. This was unfortunate because he was always personally inclined toward this position. Why he hid that remains a mystery to me.

The elimination of double punishment will always be for me a demonstration of politics at its finest, overcoming divisions and bringing together men and women of such different convictions. It's a great source of optimism, because it shows that politics in the noble sense of the word can make it possible to unblock stalemates by finding room for maneuver and unexpected cases of consensus. It shows that the French can accept that one of their political leaders can change his mind on a sensitive issue. If you are sincere, authentic, and coherent, then public opinion is much more tolerant than many think.

TAKING ACTION IN A COMPLEX SOCIETY

Many said that I supported the abolition of double punishment only to balance out my proposals on immigration. I accept that, without in any way backing away from the principled beliefs I expressed earlier. But I do think that balance is a condition for reform in societies as complex as ours. It is even a necessity.

Immigration is a very difficult issue. Pressure for migration into northern countries is exacerbated by the gaps between developed and poor countries, divergent demographic trends between north and south, increased information flows, and the development of air travel. With controlled and selective immigration, the mixing of populations constitutes a form of enrichment and a condition of renewal. But at the same time, large-scale immigration can be neither an objective in itself nor a solution to north-south problems, as those who propose the elimination of borders pretend to believe. Nor is immigration the solution to the problem of aging populations in the north, because immigrants also grow old and obviously also have the right to pensions. Immigration must be regulated if we want to avoid creating sudden imbalances both in the sending countries and in the destination countries. France has always been a country of immigration. I am well placed to know this. France's vocation is to remain open, to diversify, and to be enriched by the arrival of new peoples. But its vocation is also not to disappear or dissolve under a tidal wave of immigration.

Given how much the Jospin government weakened our immigration legislation, it was essential to take the steps necessary to control immigration better. At the same time, I did not want to go from one extreme to the other. I was never in favor of "zero immigration," which, even if it were possible, would lead the country to shrivel up. Many foreigners live legally in France and must be respected for who they are and what they bring to France. At least one in four French people has a foreign grandparent. The vast majority of the French want immigration to be better managed, but a similar majority also want France to remain faithful to the ideals of tolerance and generosity. The French want to be tough on illegal immigration, but they are uncomfortable expelling an undocumented foreigner who lives nearby. Finally, most of the French expect the State to regulate immigration, but they want this done humanely, by going after illegal immigration networks, respecting human dignity, and protecting the weak, and especially by putting into place a serious development aid policy, the only way to stabilize populations in the immigrants' homelands. All these realities—the seemingly irreconcilable needs and conflicting demands—are what make our society so complex.

Where others see contradictions and incoherence, I see complexity and complementarity. The job of a political leader is to know how to interpret these conflicting feelings and try to give them meaning, and to provide answers and perspective. The truth is that the debate is not between generous, bighearted liberals on one side and supporters of inhuman rigidity on the other. The two feelings coexist within all of us. Nor are there two versions of France, one of which is generous and open and the other stingy and impulsive. There is one France that wants all at the same time: firmness and generosity, fraternity and order, solidarity and responsibility.

By abolishing double punishment, the government sent a signal of openness and generosity to foreigners who have been living in France for a long time. It made a clear distinction between a policy of controlling migration flows and a racist and xenophobic conception of immigration. This was a balance that calmed the debate and made reform possible, and I'm convinced that many French wanted to act in just this way. It was the same way of thinking that led me to have offices of the Red Cross and ANAFE—a foreigners' rights defense association—set up at Charles de Gaulle Airport, and also to have observers present for all group expulsions of foreigners, and, more recently, to abolish the expulsion during the school year of illegal immigrant families with children in school. Because everything involving children is always highly sensitive, I asked the leading attorney Arno Klarsfeld to assess how many families were in this situation and to propose a humane policy that would enable us to deal on a case-by-case basis with our schoolchildren and their parents.

Saying that I think balance is important does not in any way mean that I was not sincere when I proposed abolishing double punishment. I was able to convince the UMP and our fellow citizens of the necessity of this reform because I was convinced of it myself. If pursuing this reform also had the effect of avoiding unwarranted criticism or misunderstanding about our immigration policy, so much the better. I would also note that balance works in both directions. It was also because I was proposing firm measures in the area of illegal immigration that the French and the members of the UMP were willing to support my reform of double punishment.

Balance is not "half of a good idea." France has pursued that sort of wishy-washy policy for far too long. Balance means two fair and strong ideas that complement each other and make reform possible.

In the area of immigration as in other areas, I am convinced that you can't in the long run be firm if you are not fair. This is an enduring obsession of mine. In a country that puts such a premium on equality, the question of justice is critical. Everyone must have an opportunity, everyone must be recognized for what he or she does and has done, and everyone must believe that he or she can succeed. I love the word "service." I don't feel the need to always couple it with "social." In my view, when you do that, you reduce its reach and weaken it. You make it almost banal.

THE NEW CLEAVAGES

Part of the reason for the growing gap between citizens and politics—or at least politics as practiced in France—no doubt comes from our inability to understand the social cleavages within France. The uncertainty of our times makes these cleavages difficult to identify. We're living in a period of transition between two worlds. We know the world we're leaving behind but can't make out the future one very clearly.

The left tries to respond to the need for solidarity that it rightly perceives in society by supporting large benefit programs and increasingly generous redistribution plans. By doing this, it fails to take into account the fact that the French people don't want individual responsibility to be neglected. The French want everyone to be protected against setbacks through measures such as the Minimum Integration Income and generous unemployment benefits. On the other hand, they are indignant about the fact that someone can live off of this generosity for years without having to go back to work.

The left considers inheritances to be the symbol of the kind of bourgeois success that it detests. It overlooks the fact that all French people want to be able to give their children the fruit of a lifetime of work without its being taxed away. They are starting to make this clear earlier and earlier. Parents want to help their children get started in life, not wait until they are fifty-five or sixty years old to leave them an inheritance when they die. I was struck by the success I had when, in 2004, I exempted certain early inheritance gifts that could be given to children and grandchildren. I knew this measure would be popular and useful; I just didn't realize how much. Eliminating inheritance taxes for all small and medium-sized estates—that is, for 90 to 95 percent of them—is not a fiscal policy. It's a family policy that crosses old political cleavages.

The right for its part has railed against the thirty-five-hour workweek for devaluing work and for undercutting the competitiveness of our companies against increasingly tough international competition. And the criticism is merited. That said, the right doubtless underestimates the desire of many of our fellow citizens to be able, at certain stages in their lives, to find a better balance between their personal and professional lives. When you're young, you're ready to work like crazy to start a family, buy a house, and get your career off the ground. Later, you look for better balance between work, leisure, and family life. Finally, once the children have grown up, some are ready to recommence a more active professional life. I think that instead of having a rigid and one-dimensional thirty-five-hour workweek and retirement at sixty imposed on them, our compatriots want more flexibility. Those who want to earn more want the freedom to work more, and all want to vary their working hours according to what stage of life they're in.

WOMEN: LIVING THREE LIVES IN ONE

Talking about women and to women should be a political priority. It should be done without being smug, demagogic, or suggestive. The objectives should be to understand, to respect, and even to be daring. We should dare to say sensitive things without being vapid. We should dare to be genuine without being vague. The life of a woman in 2007 is more difficult than the life of a man. A woman lives at least three concurrent, sometimes contradictory, and always busy lives. She lives the life of a woman, a mother, and a working person. The big issue is how to make them all fit together. The most common ambition is to avoid giving any of them up. Deep down, the real luxury is not having to choose in order to be able to have all these feelings in a single day. To make all these female aspirations possible, we have to organize society differently.

This is all the more true in that women's accession to positions of high responsibility is transforming our society. Wherever they are, women bring with them new, different, and complementary ways of doing, thinking, and acting. Yet the promotion of women is going very slowly. Women make up nearly 56 percent of our national civil servants, but only 10 percent of the top civil service posts. They constitute 80 percent of the students of the National Magistracy School but only 8 percent of top magistrates. Only 6 percent of company directors, 10 percent of CEOs, and 12 percent of members of Parliament are women. This can be explained by a number of factors, including the persistence of unequal qualifications and the existence of discrimination, whether voluntary or involuntary, but also the difficulty for women to reconcile their many roles. Many end up giving up on trying to do it all.

There should be full tax exemptions for all family spending on household help. A family is in a way a small company. It is absolutely appropriate for it to be able to fully deduct expenses, and such a deduction would in any case help boost employment. We also have to develop a system of child care at work, and make clear that it is unacceptable for women to be punished in the hiring process or the development of their career because they're expecting a child. Yet this happens all the time. I also propose to expand to all schools a practice I experimented with in Hauts-de-Seine. This was to make it possible for all pupils to take advantage of supervised study from 4:30 to 6:30 or 7:00 p.m. This would reassure women about what their children are doing between the end of school and the beginning of the evening. And it would enable the kids to arrive home with their homework already done.

We haven't found a good way to respond to the need of certain mothers to be able to vary their working time depending on the changing nature of their family constraints. Certain women would like to be able to have free time when their children are adolescents or when their parents are elderly. The answer is to think about work time over the course of a lifetime, not the course of a week.

Finally, I am struck by the fact that France is doubtless one of the few countries where it is normal to have work meetings, often the most important ones, in the evening. This leaves many women out. As in Spain, we're going to have to make efforts to start working days earlier, end them earlier, and avoid meetings after 6:00 p.m.

THE PACE OF REFORM

I don't share the view of those who believe that everything must be put into place within one hundred days of an election, as if somehow after that the electoral momentum is gone and you can't do anything without provoking the electorate ahead of the next election.

Let me first note that the risk of not getting reelected is no reason to stop working several years before the end of an electoral term. This is an approach that has already proved to be ineffective in that, since 1981, no majority has been reelected. I also think that if certain reforms are in fact easier in the first three months after an election, the inability to act during the months and years that follow results from a failure of method, as well as a lack of faith in the degree to which the French want change. I particularly think that the biggest mistake, which is common, is to undertake reforms sequentially. First you do pensions, then education, and then finally welfare or immigration. With this system, you often end up stopping after the second reform, exhausted by the battles over the first. Thus you get all the disadvantages of change without any of the advantages. In other words, you do just enough to provoke reactionary tendencies and stir up interest groups, but not enough to win over the most modern part of society.

We had this debate just after Jacques Chirac's election in spring 2002. I wanted to take on pension, health care, and education reform simultaneously during the first three months. I was sure that the momentum this would generate would make it easier for us to win support for the reforms. Chirac didn't want to do this, because of his analysis that French society is hesitant about change and that it was too risky to rush it. Ultimately, we managed to reform the retirement system, thanks to Jean-Pierre Raffarin and François Fillon, but we had to cut back on our ambitions in the area of education. As for health insurance, everyone knows that the hard work remains to be done. Unfortunately, I can only note that after the electoral shock of April 21, 2002, voters kept sending warning shots. In the regional elections of 2004, sitting governments lost in twenty out of twenty-four regions, and in the 2005 referendum on the EU constitution, the "no" camp won 55 percent of the votes.

For me, the ability to reform depends less on the first hundred days than on a frank, detailed electoral campaign and on a government whose main course of action should be to do what it said it was going to do. From this point of view, it is clear that France was hurt by the truncated debate between Jacques Chirac and Jean-Marie Le Pen in 2002. None of the big issues for French society were resolved or even addressed during the second round of the election, because the overwhelming need to block Le Pen took top priority. The successive governments under Chirac suffered from this because their mandate was not clear enough to tackle the big issues. Since the voters were not presented with the real choices in advance, it was hard to win support for the reforms.

DIVERSITY REINFORCES UNITY

My diagnosis is clear. France is suffering not from too much politics, but on the contrary from not enough. Our response to the rise in abstention, the protest vote, and despair can be found only in our will to give new meaning to politics, to give life to ideas, and to put force behind governmental action. Besides, in the second round of the 2002 presidential elections and for the 2005 referendum on the EU constitution, the French were mobilized. When something is at stake, when there's a debate, the French get excited about politics. This is the deep conviction that led me, as president of the UMP, to look for the cement of our unity in our diversity. The Socialist Party is homogeneous but divided. The UMP is diverse, but unified.

Since 1974, the date of my first membership in a political party, during the era of the UDR (Union of Democrats for the Republic), I have never belonged to any other than a Gaullist party. This is all I've known. I've never been tempted to leave. Even when I was in a minority within the Rally for the Republic (RPR) after the 1995 elections, I never thought about quitting the party.

I was always seduced by the great orators of the Gaullist era, such as Jacques Chaban-Delmas, Michel Debré, Alexandre Sangui-netti, Charles Pasqua, and later Jacques Chirac. Their speeches reso-nated with me, stimulating my strong desire to get involved. What I really liked were the collective emotions. In victory as well as in defeat, I really loved sharing my feelings with others who I felt were both so close to me and so different. I wouldn't have given this up for anything. These long years as a simple party worker enabled me to climb all the ladders and assume high responsibilities. But I have never forgotten the aspirations of the low-level party member that I used to be. I wasn't always up on the stage. For a long time I was just in the room, and this is what makes me different today, and is really my strength. I think I understand the aspirations of the public, since that's where I come from.

This experience taught me to be suspicious about the tradi-tional reflexes that threaten all political leaders. I quickly turned away from the nightingales of partisan political rhetoric who, in the name of unity, try to silence any original proposals or new ways of speaking. How many times have I heard someone call for the withdrawal of a motion, amendment, or proposal in the name of sacrosanct unity! How many times have I heard the traditional in-cantations on the discipline that must be respected lest the party fall apart!

In my role as president of the UMP, I am motivated by the op-posite principle. I am convinced that unity is not the "cause" but the "consequence." It is the consequence of an ongoing and permanent debate, without taboos, and especially without internal tension. It is precisely because the debate will have been exhaustive that the unity will be solid, respected, and shared by all.

Competition is essential in politics. Only competition makes it possible to demonstrate one's values and to select the best people. Contrary to what has often been said, it is not a cause of division. A loyal competition can be a form of unity. I was elected for the first time in 1977. Since then, that is for the past thirty years, I have gone before the electorate fifteen times, or practically once every two years. French people who have the feeling they're always seeing the same faces don't fully appreciate this. At every stage of my career, I was confronted from within my political family. I competed with Charles Pasqua, Alain Juppé, Jean-Pierre Raffarin, Dominique de Villepin—the list is quite long. I never thought there was anything unfair about this. I always accepted it as a form of initiation that taught me, that made me give my best, that stimulated me to try to improve. It obliged me to change, led me to be more nuanced, and, consequently, led me to develop my analyses and my beliefs. In politics, it is more dangerous to be named to something than to have had to conquer it. Conquest teaches you humility and shows you how precarious things can be. Being named to something be-stows power that is not necessarily deserved. It leads you to think that you know everything, when in reality you don't. Without all this competition, I wouldn't have known how to question myself and wouldn't have been able to do so.

I am sure that personal freedom makes those who have it more responsible. The opposite is also true: constraints create the most serious tensions. Far from fearing diversity, I go looking for it. I seek out opposing arguments. I believe in respecting differences. The complexity of situations and the problems a political leader must sort out can only lead him or her to a strategy of diversity and openness. The French electorate, now so diverse, demands policy options that are diverse and ambitious, but also well thought through and tolerant. Far from being a cause for concern, our ability to debate will impose respect and will widen our political space.

This reasoning holds for ideas as well as for people. In the UMP, I wanted to be surrounded by people of diverse personalities and opinions. With so many different types of people, political journalists wonder about our ability to propose a coherent project to the French people. They say we'll have to arbitrate over tough choices. My goal is not to arbitrate. It is to decide priorities, to reconcile, and to unite. A balanced program will emerge from our wealth of convictions.

I see France as a kaleidoscope. It has many faces, many facets, and a variety of aspirations, and this is all to the good. The question is whether the final image we're able to make together will be a success. Will it be dull, banal, common, boring, and insipid? Or will it on the contrary be harmonious, brilliant, and fascinating to behold?

JACQUES CHIRAC

I am not alone in believing that the debates I had with Jacques Chirac enabled us to turn our majority into the forum for dialogue in which the French people can recognize their diverse aspirations. I admit it was not always easy. And sometimes it may have gone too far. I'm perfectly prepared to admit my share of responsibility for some of the clashes that perhaps needn't have been so sharp. Some will say that this is a matter of temperament. But you could also say that it's a matter of candor. Those who know me know that I do not like to lie, either to others or to myself. I say what I think. I do what I say. This is not always an advantage, but that's the way it is.

Therefore I'd like to tell you about my relationship with Chi-

rac. So much has been written on this already. But I don't recognize
the true nature of our relationship in the way it's usually described.
First, we spoke to each other much more often than people think.
During the years when I was finance or interior minister, we had
at least one private meeting per week. On top of that there were
the cabinet meetings, at which I was seated just to his right, and
numerous other meetings on all sorts of subjects. Our common pen-
chant for people and for discussion means that even in the periods
when tension was highest, we were incapable of not speaking to
each other. All these interactions were not enough to overcome the
disagreements, but they at least made it possible to avoid misunderstandings. And direct contact avoids the problem of entourages getting involved and making an already complicated situation worse.

I have often heard people characterize our relationship as one
of "inexplicable or visceral hatred." I cannot and do not wish to
put words in the mouth of Jacques Chirac. But I can speak in my
own name. "Hatred" is an entirely foreign sentiment to me. I don't
hate anybody. And at my age, moreover, it would be unreasonable
to be so puerile. I can even say, without exaggerating, that I admire Jacques Chirac's qualities. I admire his energy, his tenacity,
his force of character when faced with adversity, and his ability to
seem—and therefore to be—friendly. These are characteristics you
don't come across very often. His career has also been exceptional.
He was a deputy from Corrèze, mayor of Paris, minister several
times, twice prime minister, twice president of the Republic. You
don't attain such longevity without possessing and developing an
exceptional temperament. This is undeniable. I don't need to flatter
him and I'm not doing so. By writing this, I have no motive other
than to explain as accurately as possible the way I see it.

Now of course, we have also had our disagreements. I supported Édouard Balladur for president in 1995. Given the great qualities of the former prime minister, I don't see the slightest reason to
apologize for that. Jacques Chirac himself in his day opted to back
Valéry Giscard d'Estaing instead of the candidate from his political family, Jacques Chaban-Delmas. I believe he's been around long

enough not to hold it against me that I did just what he did at my age. On the other hand, what is true—and he's told me this several times—is that he was sometimes irritated by my desire to remain independent and my tendency not to follow instructions that don't correspond to my convictions. Maybe it's because this temperament reminded him of somebody else!

Finally, there were substantive disagreements between us. They were never insurmountable and they didn't prevent us from governing, but they explain that he is who he is, and that I am who I am.

The first has to do with our political positioning. He was always reluctant to consider himself part of the "republican right." I don't feel the same way. I see doing so as a way of providing clarity that is necessary, salutary, and useful. He sees the president of the Fifth Republic as an arbiter who brings people together and resolves disputes. I see the president as a leader who decides and takes responsibility for his decisions. He thinks France is fragile and resistant to change. I think France is impatient, exasperated by delays, and eager for profound change. He believes more in the quality of people than in the strength of their programs. I don't think you can win elections without saying exactly what you can do and are going to do. I believe in the modernity of ideas. In my view, not everything comes down to empathy.

We're not irritated by the same things. He gets irritated with liberalism, Americans, certain CEOs, and people who disagree with him about Europe, whom he quickly comes to see as irresponsible incompetents. I get irritated by the lack of steadfastness, by hesitation, by unkept promises, by the refusal to see France as it is, and by conventional wisdom. There's even a big difference in the way we give speeches. He belts out addresses that have been carefully prepared with his inner circle. He knows how to get people motivated. I meticulously write my own speeches, and seek more to convince people than to fire them up.

It's no secret that he didn't want me to run for the presidency of the UMP, and I went against this presidential diktat. Yet I think that I can honestly say that every time we faced a difficult situation, I was able to count on the confidence of Jacques Chirac. Professionally, or I might even say "technically," he always gave me lots of freedom. "Politically" is another issue. In crisis management—on issues such as the suburbs or the CPE—I often found myself sharing his analysis and I appreciated, each time, the confidence he showed in me. As for the conclusions to draw from these crises, that's different. Our differences were as much about substance as style.

So that's my relationship with Jacques Chirac: deeper, more complicated, and more direct than most people think. I have often wondered what one had to do to deserve political confidence while refusing to be submissive. A lot of people who would call themselves Jacques Chirac's friends have caused him a lot more trouble than I ever will. Yet I'm not on the list of friends. I've accepted this once and for all. It's a given, and I've decided to live with it.

CHAPTER 9

A Clean Break

I first spoke of making a "clean break" at the UMP summer meetings in September 2005. I still remember perfectly the disturbed look on the faces of many of my friends and some of my advisers when I started to invoke this theme. I was expecting this reaction. What had become obvious to me after having thought a lot about it was for them a novelty and a surprise.

Everyone had comments. The idea of a clean break was too violent, not polite enough, even a cause for anxiety. It was going to upset people, when I should be reassuring them. It was going to lose us the vote of the traditional right. It was going to confuse elderly voters. Why did I yet again fail to speak like everybody else? I could have used words such as "change," "alternation," or "reform," which meant the same thing but didn't have the same disadvantages.

In truth, I made a deliberate choice to talk about making a clean break. Traditional alternation between left and right had left people disappointed, the word "change" had lost its significance because reforms were never followed up, and the word "reform" itself had become meaningless because it was overused. As for risks, who can hope to win the presidential election without taking any? I believe in more generous political action and total commitment. It wasn't long before the facts started to come in on my side—because in the weeks after my speech, all of my opponents started to go out of their way, in one manner or another, to associate themselves with the notion of making a clean break.

BREAKING WITH WHAT WEAKENS US

The clean break I want to see is not a break with the France that we love. It is not a break with our ideals, our values of solidarity, our conception of the State, our tradition of openness, or our ambition—so audacious but now so natural—to influence the destiny of the world. The break we need is a break with the old ways of doing things. It's a break with the way we've approached politics for years. It's a break with this inertia that in fact clashes with our ideas, our values, and this "certain idea of France." Finally, it's a break—and this is the main goal—with the despair that saps our inner strength and opens up the floodgates to all sorts of extremism.

I can already hear those who will bring up my old loyalties, my past commitments, my long participation in this country's politics. I'm not walking away from any of that. But what would be catastrophic would be to not have changed anything in thirty years. Would it be better to be the same as ten or twenty years before—with the same impatience, the same ignorance, the same excesses, the same insufficiencies, the same diagnosis, and the same program? Would it be preferable not to have drawn any conclusions from my failures, or from the three electoral shocks that have hit our country since 2002? I am not the same as I was thirty years ago. I will not be the same twenty years from now, if God allows me to live that long. And that's a good thing!

Unlike many young people of my generation, I was never seduced by the leftist utopia of the 1970s or by the nationalism of the far-right groups on the other side. My core beliefs—the values on which I built my public and private life—are faithful to what I took from the reading I did and the experiences I had at that age when you start to think for yourself and your ideas come together. They've never changed since. They have gotten richer, more nuanced, and probably simpler, because life manages to flush out and destroy the delusions and wild ideas that seem so compelling when you're young. At least I hope so. But I'm not walking away from any of this, just as I don't regret any of the important commitments I've made. This is for me a source of pride.

This does not prevent me from holding up these ideas and my past actions against the reality of France today. This is in fact my duty. And a lot of what I see leads me to think it really is a clean break that we need. This includes my government experience between 2002 and 2007—during which I tried to apply the lessons I drew from the 2002 presidential election, the results of the elections we've had since then, the evolution of the world, France's role in it, and the way people see France.

There's nothing disturbing about a break with the past. It's not a rejection of past convictions but the desire for a change in approach and, most of all, the energy to do it.

ENDING THE LIES

For twenty-five years, our country's dynamism has been stifled by the juxtaposition of three lies. This is the first thing we've got to break with.

The first lie is pretending that it is fairer to share nonexistent wealth among everyone than it is to help individuals create wealth.

The French Socialists' specialty is to distribute wealth that doesn't exist. They raise the minimum wage, social benefits, and the number of civil servants, and they build low-cost housing. This is a seductive approach—how can anyone resist it? No one accepts inequalities. No one wants a less socially conscious Republic. No one claims that you can live well today on the Minimum Integration Income or the Specific Solidarity Benefit or that it's easy for a family on the minimum wage to make ends meet. I am also as aware

as anyone that it's necessary to improve the lot of all who are struggling. It's not only a humanitarian issue but in the interest of society
as a whole. It is the condition of social balance and social cohesion.

But before you can distribute wealth, you have to create it.
And the facts are there: France has been creating less and less
wealth for thirty years. The French Socialists refuse to admit what
British, German, and Scandinavian Socialists have understood for
a long time. After the illusion of the "youth-jobs" and the thirty-
five-hour workweek, they come up with the illusion of a minimum
wage at €1,500 per month before taxes. Acting in this authoritar-
ian and statist way, disconnected from real productivity gains, they
are trading the hope of a more comfortable life for a certain risk of
real job insecurity. Raising the minimum wage without taking economic realities into account is to condemn companies to cut back on
production, scale down hiring, and shed thousands of jobs. What a
few will gain, many more will lose. Just as with the thirty-five-hour
workweek, the numbers won't add up.

I also challenge the idea that this measure is fair. Simply di-
recting that the minimum wage be raised to €1,500 is to demand
of the French that they give up any personal ambition. This step
would absorb all room for maneuvering that companies have in
the area of compensation policy. It would affect only those French
who receive the minimum wage, and not those who are paid a bit
more than that, and whose wages have not risen for years. No longer
would people be compensated on the basis of their efforts and their
length of service to the company; rather, everyone would get the
same thing.

To me, buying power is a key issue. It can and must be en-
hanced. But it's not by undercutting companies' competitiveness or
creating employment that we're going to do it. It's through the dynamism of our economy and full employment. Full employment
is possible. As strange as it might sound, the key is to be found in
work. It's by working more that we'll create more jobs, since each
person's work creates work for others. A senior citizen who starts
working again, a young person who signs his or her first contract, a

woman who goes from working part-time to full-time, job cuts that are avoided—all of these things lead not only to more buying power among the workers concerned but more jobs for the whole economy thanks to the consumption and activity this extra work creates. This is why I think that our priority should be to provide incentives for work and to compensate it. We can do this by allowing our companies to succeed and thereby raise the salaries of all workers and by exempting overtime hours from all income and payroll taxes—this is the only way to overcome the big mistake of the thirty-five-hour workweek. My view is that it's through work that we will create jobs and buying power for everyone.

The second lie is about our national debt. It has risen from 20 percent of GDP in 1980 to 66 percent today.

Certain so-called experts and political leaders would have the French people believe that the public debt is not a problem, that national debt is different from private debt, that France will always be able to find lenders to lend to it at reasonable rates, and that its rank among nations will always allow it to remain among the most secure borrowers. The same people go so far as to affirm that the debt will one day be canceled. I don't see myself as dogmatic in any way. I distanced myself a long time ago from the classical economic thinking—from the "strong franc" or "strong euro" policy no matter what the cost—that has been responsible for slowing our economic growth. But it would be crazy to think that France can keep piling up debt without risk.

France has debts toward foreign and domestic investors.

None of them plan to give what they're owed back as a gift. Whatever debt we don't pay back today will have to be paid back by our children tomorrow. France is not protected against a possible downgrading of its creditworthiness, which would raise the price at which it borrows money. Nor is it protected against a general rise in global interest rates, which has already started. The interest paid on the national debt is already equivalent to the entirety of France's revenue from income taxes, and a greater amount than France spends on national defense. This is not small change.

What's worse, our debt is being used to finance not future spending—which boosts future growth and well-being—but current spending. We spend less per student in our universities than do other countries. We spend too little on research. While the price of oil has reached high levels and the possibility of exhausting our petroleum reserves is becoming a reality, we haven't started to adjust our transportation infrastructure accordingly, by promoting walking and biking, public transport, and rail travel. Nor have we done enough on renewable energy. Similarly, the equipment necessary to support dependent elderly people is insufficient. I'm not making debt and debt relief into some kind of dogma; they are just a question of good sense and respect for future generations.

CIVIL SERVANTS: ACTORS FOR CHANGE

For years, many civil servants have turned to the left out of despair about a right that always seemed critical and sometimes even lacking in respect. In fact, if our view on civil servants consists only of seeing them as "excess fat," and if our program consists only of reducing their numbers, there's little chance that any government worker—even those who share our thinking—will come over to our side. I think we need to start replacing only one out of every two civil servants who retire. The State needs to spend less. Salaries and pensions in the public sector represent 45 percent of the national budget. Reducing these costs is an unavoidable necessity that can't be put off.

Does this therefore mean that we can't talk positively about the world of the government sector? Not at all. The first thing to do is to devote 50 percent of the gains from reducing the number of workers to salary increases. There will be fewer workers, but they'll be better paid. This is a fair plan. Then we have to break with the logic of exclusively relying on competitive exams to determine promotions, which sometimes leads to rewarding those who focus on studying for tests more than those who are devoted to their service.

Fifty percent of jobs open to those who pass the competitive internal exams should be set aside to reward experience and performance. This would be better than having promotions determined by increasingly difficult theoretical tests. Such a program would do a lot to facilitate social promotion and make careers more fluid. Finally, we should encourage internal mobility within government service. Having to take a test to change from being an administrative officer at the Finance Ministry to being an administrative officer at the Interior Ministry is an example of useless rigidity. All civil servants suffer from this, especially when they want to change regions. The right to training for a new profession and the right to work in a new area should be guaranteed. Civil servants who are better paid, better trained, better treated, more highly regarded, and less numerous—this seems to me to be a useful program for France. It should be attractive for the civil servants, whose devotion and competence in my view deserve more respect than is often granted them. I would add that in the public sector, the thirty-five-hour law is a great burden. This law muddled up our public services. We have to allow the civil servants who want to be free from this rule to be able to work and to earn more.

REGULATED LIBERALISM

A final untruth that I want us to have a real debate about is that of antiliberalism. I understand very well why someone might be antiliberal. Conventional wisdom bothers me so much that I would never deny anyone the right to think what he or she wants. But in the end, you have to be clear on what you're talking about.

Let me first note that people in France always talk about "ultraliberalism," but never "ultra-socialism." I believe the French think clearly enough to see the difference between a liberal conception of society and the economy on one hand and a simple-minded and dogmatic ultraliberalism.

Being liberal doesn't prevent me from thinking that the liberal economy needs regulation, norms, and constraints in order to serve citizens rather than having citizens serve it.

Such measures include labor laws, minimum wages, union laws, worker representation, consumer rights, and antitrust policies.

I am convinced that, in certain sectors such as culture and sports, specific rules must be applied to levels of pay. I also think that public services are necessary, because some goods and services are unique or so critical that they cannot be subject to the laws of the market.

Even more, I think the State has certain responsibilities, and in particular that it must have industrial and research policies. This is what I did when I defended Alstom, when I supported the Sanofi-Aventis merger, when I created competitiveness centers, and it's what I'm doing now in creating centers of rural excellence. I think that we went too far in giving the European Central Bank control over monetary policy without setting out the conditions for an institutional debate with political leaders. Inflation used to be the major problem for European economies. Today it's mainly growth, and related to that, unemployment. Just as in the United States, ministers of economics and finance need to have regular contact with the European Central Bank to create together a monetary policy that respects the role and prerogatives of each.

Finally, I don't see in what way liberalism is inconsistent with social policy. On the contrary. The role of the economy is to produce wealth. The role of politics is to share wealth equitably.

But what I really want to stress is that the market economy, which is really just another name for liberalism, produces more wealth than communism, which is really just another name for antiliberalism. The lesson of history has unfortunately turned out to be revealing on this point. No ambiguity is possible. Communism has led to misery and dictatorship everywhere it's been tried. Freedom has never been counterproductive in terms of democracy and growth. I do not understand how anyone could think, or even less say, that "liberalism would be as disastrous as communism."

The German Social Democratic Party recognized the soundness of the market economy in 1959. Yet this doesn't stop the French Socialist Party, in 2007, from flirting with the extreme-left, Trotskyite, and other antiglobalization activists, and to seriously envision governing with them. Seventeen years after the fall of the Berlin Wall, this is a strange reading of history. It leads our country nowhere and it is a rather disrespectful posture toward the millions of victims of communism. This may be disturbing but it's the truth that must be spoken. And I wish the French Socialists would show the same intransigence toward the extreme left that the right shows to the extreme right. The ideas, principles, and conceptions of these people are simply incompatible with the values of our country.

BREAKING WITH FRENCH BACKWARDNESS

Then we must break with inertia. I'm not going to list here all the reforms I think are necessary. That's not the purpose of this book. I only want to note that by putting off action, France has fallen behind and handcuffed itself in a number of areas. There comes a time when it's better to rebuild from scratch on the basis of first principles than to continue to pile up new systems on top of old ones.

In October 2005, then Prime Minister Villepin gave his view on the idea of making a clean break by noting that "General de Gaulle wanted to get things moving through continuity, not by breaking with the past." This is not my view. The magnitude and number of reforms undertaken by General de Gaulle, just after World War II and after he came back to power in 1958, mean that Gaullism itself was a clean break. Thus General de Gaulle changed French institutions, first by authorizing the vote for women, then by overseeing the development of the Fifth Republic Constitution. He transformed economic policy through the planning process and nationalizations. He gave France energy and transportation policies by creating the French Coal Board and the Atomic Energy Commission in 1945, the General Petroleum Union in 1958, and Air

France. He created the welfare system, the general statute on public service, the National Administration School (ENA), profit sharing and worker participation, and—along with André Malraux—cultural policy. He changed the currency (with the creation of the new franc), invented the modern hospital, and reformed the justice and tax systems. He brought agriculture into a new world, that of modernization and the common agricultural policy. And of course, he put an end to colonialism and gave France completely new options in foreign and defense policy, including the creation of the Franco-German couple with the 1963 Elysée Treaty, the withdrawal from NATO's integrated military commands in 1966, nuclear deterrence, and the Phnom Penh speech in 1966. It will take a while to convince me that this was continuity!

You don't want to change everything just for the sake of change. But you shouldn't hesitate to take things to the limit when it's necessary. I'd like to give a few examples.

The French fiscal system has a number of disadvantages, even beyond the fact that taxes are too high. First, it's extremely complicated. It's a tangle of taxes, niches, exemptions, exceptions to the exemptions, and so on. No one can find his way through it anymore. Our corporate taxes are too high compared with those of our competitors, particularly because of the professional tax, which everybody acknowledges has become economic nonsense. Our personal taxes have all the disadvantages and none of the advantages of those used by our European partners. The income tax is paid by only half of households, and those who pay feel that they're the only ones contributing to the common pot. But the weight of the value-added tax, the Generalized Social Contribution (CSG), and the property tax—which is flagrantly unjust—all lead to a household tax system that in the end is not equitable. The taxes on workers continue to discourage work. Finally, we have little in the way of environmental taxes.

A clean break would mean having the courage and the energy to remake our fiscal system from scratch. It should be simplified. It should be fairer, more effective, and less of a deterrent to work, risk, and initiative. It should be modernized, by moving toward collection at the source for the income tax. Direct household taxes should be brought together into a single tax, on income, with an additional bracket for the "solidarity tax" on large fortunes. There should be a different local tax at every level of government, rather than having each locality take a small percentage of lots of different taxes. The proliferation of taxes, various charges, and other assessments should be stopped, and replaced with a few taxes that are clearly identified and intelligently conceived to balance yield and justice.

THE IMPORTANCE OF KNOWLEDGE

I want to raise the issue of our higher education and research systems. There are critical for the destiny of our country and for its place in the world of the future. Every day, globalization increasingly pits national higher education and research systems into competition with one another. In the era of the knowledge and information society, the performance of our system will be more critical for the future than ever before. This system will determine how well we master the most advanced science and technology. It will shape our wealth-creation potential, and therefore our quality of life and that of our children, over the long term. It will be a critical factor in the competitiveness of our economy and the attractiveness of our territory.

In the face of this considerable challenge, I regret to have to note that our system does not meet the top international standards and that it is showing serious signs of weakness. Our research effort has fallen since the mid-1990s to stagnate at around 2.2 percent of GDP, a level well behind other industrial powers, including not only the United States, Japan, and Germany, but also Sweden and South Korea. Even China now spends more on research than we do as a share of overall world spending. As for the scientific impact of our research work, it is also falling if judged by our falling share of world publications, scientific citations, and awards. Higher education is hardly better off. Spending per university student in France is 20 percent less than the OECD average. Even more atypically, it is even less than what we spend on each elementary and high school student. Only 37 percent of each age group in France undertakes higher education, compared with 64 percent in the United States and more than 70 percent in the Scandinavian countries. Even worse, the failure rate in the first years at our universities is among the highest in the world. More than half the students registered fail to obtain any degree at all. This is a great disappointment for thousands of young French people.

There is nothing inevitable about this, however. France is not lacking in either assets or talent. It just needs to create the conditions for these to come out.

Of course, it is urgent that the resources devoted to higher education and research be increased. But that will not be enough if we do not also undertake far-reaching reform of the organization and functioning of our system, whose basic structures have not changed for sixty years. Our universities are too small and not autonomous enough. They are at the periphery of a research effort that is dominated by big organizations that absorb the lion's share of the money. They have to compete with the Grandes Écoles and postsecondary preparatory schools, which can choose their students yet don't train enough of them, and which are unknown abroad. Our universities do not have strong enough links to their economic environment. In sum, they are not well placed to compete with universities abroad, and this makes it difficult for France to attract and keep the most promising students and researchers. But universities are the crucibles of innovative societies everywhere in the world. Universities are where science, youth, and the world of economics meet, and where research laboratories and innovative high-tech companies gather.

We should promote the development of powerful and autonomous universities that can play a central role in the training of our elites and in the research effort. The autonomy of universities is a critical element in reforming our higher education system. They must have the means and freedom to govern themselves and to manage themselves efficiently. This applies to the definition and implementation of their strategic plan, their recruitment, the assignment and salaries of their personnel, and the diversification of their resources.

I think it's absolutely necessary to increase, and not ration, the number of degrees given in higher education. That's why I am firmly committed to the principle of guaranteeing a place at the university for any high school graduate who wants one. But we must be able to raise the question of selectivity at universities—as practiced by all our European neighbors—without creating a drama. I don't think our current selection process is in students' interest, with potential failure on entry exams or unemployment and lack of qualifications at the end.

For young people who want to develop their general knowledge and to have additional tools before making a definitive choice of area of study, we could create a first university year focused on general instruction.

This, I would say, would be a clean break for our higher education and research systems. Such a revolution could not be immediately undertaken by all eighty-five French universities, but it must first be led by those who have the will and the capacity to take responsibility for this change. But it's also a program that responds to the challenges of our age and the world that is emerging. It's a program whose goal is to offer the best to our teachers, researchers, and students to strengthen the future and rank of our country. When China and India are producing millions of engineers and researchers, France can no longer afford the luxury of training, in certain areas, many more students than there are jobs that have been created in these professions since 1945.

RECONCILING SCHOOL WITH SOCIAL PROGRESS

I'd like finally to raise what for me would be a clean break in education policy. For thirty years, the left and the right have looked at school through ideological lenses that are no longer relevant to the expectations and needs of educational institutions. On one side, you have the disproportionate increase in means, the reign of pedagogy, and electoral corporatism. On the other, you have efforts to cut budgets, excessive emphasis on apprenticeships, and denigration of the teaching profession. All these stances, because they're such caricatures, have left schools to fend for themselves, and families disappointed. The French people remain today profoundly attached to their "republican" system of secular, free, and coeducational schools. But they are more than ever convinced that there is no social progress without progress at schools, and schools want nothing more than to help. Let's share with them the challenges we want them to help us meet, and show confidence in their ability to do so.

In a society in which knowledge and expertise are going to be increasingly critical, I do not think that we should reduce spending on education. In fact we need to do the exact opposite. But don't count on me to promise that we'll increase spending, because that would be lying to the French people. I think we can and we must do better with what we've got.

All studies confirm that the most important factor in success for pupils is the pedagogical quality of the teachers. Unfortunately, for a number of years, we've been looking for the philosopher's stone in the Training Institutes for Primary School Teachers (IUFM) and in teaching manuals. It is to be found in experience. No administrative guidance can replace the experience of a teacher who has for years taught reading to dozens of different children. I thus support allowing each teacher to choose his or her pedagogical method. The education system will rate teachers according to the results of the students, not according to the degree to which they apply pedagogical methods imposed from above.

I also think that the importance of the teaching profession for our students and for our country is not sufficiently reflected in teachers' paychecks. Promising to raise the pay of our 900,000 teachers would be easy but it would also be fallacious. I will not do that. We just don't have the means. However, I do think that we must allow teachers, especially younger ones, to work more in order to be better paid if that's what they want to do. They could take charge of the new supervised study periods that I'm proposing to improve the daily lives of women. Similarly, the teachers who would agree to stay at school in between their classes, to ensure an individual presence among the pupils, would get bonuses. Everyone would benefit from these measures.

Finally, facing the vast and complex world of education, the tendency—especially on the right—has been to want to impose uniform, radical, and definitive reforms by waving a magic wand. But these reforms have never succeeded, because consensus is impossible and the capacity to implement them has been nonexistent. I would like for us to have a free-choice culture. Let each family choose what is best for its child and let each school adopt the pedagogical project that inspires it. For a number of years, for example, people have talked about the idea of importing into France the German model of education, which lets children out in the early afternoon so that they can play sports and do cultural or group activities until the evening. I believe strongly in the role of these three kinds of activity in the development of qualities that are not emphasized by academic institutions yet are critical for adult and especially professional life. These activities give you a better feeling for other people, charisma, team spirit, and creativity. I also believe that sports, the arts, and social activities can enable children who have academic difficulties to get interested in making an effort and going to school. What matters is not being good at everything but being good at something to develop self-confidence. In fact I would have liked to go to school in the German system. But it would be unrealistic and contrary to my conception of individual liberty in a society such as ours to impose this system everywhere. What I do think is possible, and modern, is that families in every city get the choice. The pupils who would like to try the "part-time" system for sports, culture, and social activities should be able to do so. Those who don't want to can stay in "traditional" school.

BREAKING WITH RIGIDITIES

Making a clean break also means having the courage to put an end to the ideological prejudices and formal principles that box France into a corner. Let me take affirmative action as an example.

In 2002, at the first meeting I had with all of France's prefects, I was surprised to notice that not a single one of them was of North African origin or black. This was a step backward from the time of General de Gaulle, when a governmental decree of October 29, 1958, required the government to set aside 10 percent of category A and B jobs in public service for French Muslims.

This sad reality has a number of different causes, including social stratification and the ghettoization of neighborhoods, which is sustained by uncontrolled immigration and the flight of all those who can leave. Another important factor is the role of social codes. These codes make it easier for the descendants of immigrants to get jobs in law or medicine, where hiring is based on objective criteria, than to enter preparatory schools for the Grandes Écoles or political studies institutes, where the tendency to follow social conventions has gotten worse in recent years. Finally there is discrimination. For some people it is deliberate, and for others it is subconscious and involuntary, produced by the system and old habits. I am convinced, for example, that a good part of the glass ceiling that a lot of women and descendants of immigrants run up against can be explained simply by the fact that people don't "see" them—they just don't think of them for certain positions. A minister looking for a chief of staff inevitably thinks of a white man of around fifty who studied at ENA—simply because for thirty years he's never seen anything else. If you made him look at several candidates with different profiles, he would have to imagine other people in the job, and he might realize that they could work well.

Making an effort to consider different profiles for each nomination is a form of positive discrimination, activism that is far removed from the idea of quotas. And let the best person win! That's why I supported the nomination of a "Muslim prefect." And it's why

I later supported people from immigrant communities or overseas territories for the posts of equal-opportunity prefects that were created after the suburbs crisis of fall 2005. If I hadn't obliged the Interior Ministry to look beyond the seventeen subprefects who were waiting their turn to become prefect, we would still be in the same place, and we would still be there ten years from now as well.

My proposal set off a huge controversy. The president of the Republic called it "inappropriate" all the way from Tunisia. I never understood in what way the desire to want to diversify the hiring of our elites would be against the republican ideal. What should offend the Republic is the idea that someone's prospects for promotion should depend on the color of his skin or what his name is—not that someone should want to end this injustice. It was at this time that I realized that the French love to denounce inequities without wanting to do what's necessary to overcome them.

Why this expression "Muslim prefect"? Some wanted to see in this a religious connotation. Thus I was accused of wanting to categorize part of the population according only to the criterion of religion. This criticism is absurd. When you talk about French Jews, you are referring not to those who go to synagogue but rather to those who see themselves as part of an identity that is more cultural than religious. Well, it's the same for the expression "Muslim," which is not designed to confine anybody to a mosque but which gives a name to those of our compatriots for whom Islam is part of their identity. This is, moreover, the expression that General de Gaulle used in various documents setting aside certain government jobs for them. Anyway, what else would you call them? Would you call them "Arabs"? That would have an ethnic connotation, which in addition wouldn't even be accurate for some of them, such as Berbers, Turks, and sub-Saharan Africans. The same would apply to the idea of calling them *"Maghrébins,"* or "North Africans," which would apply only to a few countries.

We could call them "French descended from immigrants"—but in this case I would belong to the group, and no one describes me that way. Then there's "visible minorities," but that would mean defining an identity solely on the basis of its being a minority one. You see, in fact, the controversy was purely political, and it was meaningless.

The ban on keeping statistics that take account of the origin of French people is equally meaningless. Our statistics recognize only two categories of people—French and foreign. All the others can be calculated only on the basis of polling—and even this only under restrictive conditions—or indirectly by extrapolating from someone's birthplace or that of his or her parents. It is thus forbidden in France to calculate the number of French who are of North African origin, Turkish origin, Chinese origin, or black. It is thus impossible to know if one group or another is more or less affected by unemployment, failure in school, certain health problems, or housing problems, or if a group is well represented in certain professions or areas of study or absent from others. To prohibit this research is to prohibit attempts to measure the diversity of France and thus to refuse to do anything about it. Once again we're favoring form at the cost of resolving a substantive problem. As for the technical difficulties involved in getting these statistics, which lead some to fear a return to the sort of sordid ethnic definitions that we don't want to see again, they don't really matter. All the other countries have managed to resolve them, even if only by basing their data on the voluntary declarations of the people concerned.

I've noticed that certain measures that in 2003 were "inappropriate," "contrary to our traditions," "against the French model of integration," and "opening the door to Anglo-Saxon-style communitarianism" have now become almost trendy. Today countless magazines, conferences, and speeches suggest affirmative action as the way to remedy problems of social integration of people living in troubled neighborhoods. A High Authority for the Fight Against Discrimination and for Equality has been created, and its powers have been widened. A special high school will be set up in Seine-Saint-Denis to give students an opportunity to get onto our most prestigious educational tracks. Partnerships with Grandes Écoles and tutoring programs are being expanded, and every company is drafting its own diversity plan.

I'm not going to complain. I only regret that people felt it necessary to begin by caricaturing my position on this subject in ways designed to discredit me. Actually, what is most reprehensible is the effort to discredit these ideas. Acting in this area should be a top priority. France must overcome the obstacles that ongoing discrimination places in the way of social success for people from immigrant families. If it does not do this in the next five years, resentment will be impossible to overcome, and we'll suffer from growing ethnic divisions. Thus we really do need a clean break.

We have to undertake major educational reforms, not by grouping all students in difficulty in the same schools (as is the case for ZEPs), but by giving each of these children the support necessary to succeed. We have to enable children descended from immigrant families to get on the fast track toward secondary education by fighting against social conventions and low expectations. Too many children give up, simply because they think it's not for them. There are fewer children of laborers and blue-collar workers in the Grandes Écoles today than in the 1950s. You can hardly call this progress. To get where we need to be, we have got to expand experiments such as those undertaken by Sciences-Po and ESSEC. These top schools have partnership programs with certain high schools in disadvantaged neighborhoods designed to help the children there with the entry process. The methods vary and are the subject of debate, but that's not the most important issue at this stage. More generally, every school should send the school records of its best students to the postsecondary preparatory schools (currently half of schools do not send a single student file to these schools). And these preparatory schools should be required to set aside a certain number of places for students from disadvantaged high schools. Special training sessions should be given to help prepare for the entry exams for public service. Finally, the government, business, the media, political parties, and NGOs must make an effort to promote diversity in hiring both blue- and white-collar workers. It is urgent not only to denounce the injustices but to give ourselves the means to reduce them. Let's see more energy in our actions and less outrage in our words.

One way to do this would be to take a cue from the American NGO Big Brothers, Big Sisters, which has for more than a century helped thousands of less well-off Americans advance through work, merit, and competence. France, which has been looking for opportunities to create social linkages to the point of thinking about reestablishing military service, could encourage students and young workers to commit to tutoring programs for disadvantaged pupils and their families. I have long thought that generosity and benevolence are not adequately rewarded in our country. Someone who devotes time to others creates loads of administrative problems for himself. Does he have the right diploma? Is he following the regulations? Are his reimbursements in compliance with Circular 24-7B-47E of the tax authority? Or who knows what else? Benevolence and commitment must on the contrary be rewarded by recognition and acknowledged on CVs. College credit should be given for volunteer work, which should also count as work experience. We could even think of giving it tax advantages. We do this for monetary gifts. Why not do it for gifts of someone's time, since it's the time people give to each other that is most lacking in our society?

CHAPTER 10

Fixing the French Economy

As France enters the twenty-first century, it faces four key challenges. The first is the challenge of globalization, which requires France constantly to measure itself against others and not just against itself. The second challenge is integration, that of maintaining a common culture, shared values, and equality of opportunity among its citizens. The third challenge is demographic, that of coping with an aging population. The fourth challenge is environmental, with future generations obliged to deal with the consequences of the changing of the earth's climate.

France has not done enough to meet these challenges. Indeed, given this failure, it is surprising that France is as productive and influential as it is. The country's current condition is all the more unacceptable in light of what it has been able to accomplish in the face of such great difficulties. France is a rich country that wastes its resources. It wastes its human capital through mass unemployment, brain drain, youth inactivity, early retirement, and the thirty-five-hour workweek. The State is living above its means, and France is living below its resources and capabilities. Its economic growth is slower than it should be.

France is a rich country that is getting poorer, is losing self-confidence, and is causing its youth to lose hope. Growing poverty is hidden behind the wealth of a few, and not enough is invested in the future. This relative decline is not inevitable, and neither is the identity crisis it causes. And decline does not result from who we are. Our history, culture, values, and language are not handicaps. Nor is the French people's worldview; its conception of the nation, the State, liberty, and equality; or its "Cartesian" way of thinking. Indeed, these are our strengths.

There is no single cause of our difficulties. But in my view the main cause is a crisis of values. Our decline is due most of all to a moral crisis that is shaking our sense of national identity and authority as well as the sense of solidarity that is a core value of the French Republic. At the heart of this crisis is the devaluation of work. France's rebirth will result from work. Everything becomes possible if work becomes a respected value once again and if working France regains hope.

I want to put work at the heart of politics and society. I don't believe in the utopian ideas that suggest that wealth can be created from nothing. Such thinking has always proved costly to those who have allowed themselves to be seduced by it. I do not believe in the fantasy of a new economy based on the end of work. I think everything must be earned. I believe in the moral, civic, economic, and social value of work. The world is working more. France is working less. That's our problem.

The French left broke ranks a long time ago—in deed if not in word—with the legacy of Jean Jaurès and Léon Blum. It has distanced itself from workers whom it has essentially betrayed because it no longer understands them. The French left proposes a minimalist society. It stands for minimalist social rights, salaries, education, security, respect, effort, success, property ownership, profits, duties, and work. This left that is no longer really the left at all even wants minimalist politics and a minimum of debate.

I want the opposite—a maximalist society. I want maximum salaries, buying power, culture, security, success, property ownership, rights and duties, respect, effort, and work. This is because I think these are society's real values. During the French election campaign, I tried to show that the election was not only about two different conceptions of politics or economics but also about two conceptions of humanity. It was about two different conceptions of what France is capable of.

This is not a matter of preaching to the French. They have made enough sacrifices over the past decades that they don't need lessons from anybody. They are among the best and most productive workers in the world. French workers, whether in the public or private sector, are not responsible for unemployment, exclusion, public debt, rising taxes, or eroding buying power. French workers are the victims.

France is not being dragged down by laziness or by an unwillingness to take risks. It is being dragged down by politicians who have all too often gotten their priorities and values backward.

The deficits that are paralyzing the French State are less the cause of our social divisions than our social divisions are the cause of our rising deficits. We are now paying the price of these divisions, which we have never really tried to reduce by addressing their root causes. Our deficits result from slow economic growth, unemployment, and the fact that many French people who could work are excluded from the labor market. They also result from the cost of resources wasted on the huge amount of public sector hiring designed to compensate for the weakness of private sector job-creation, wasted subsidies, and clientelism.

The fiscal downturn has increased the cost of labor and diminished buying power. This, in turn, has undercut our competitiveness and reduced demand, raised unemployment and slowed growth—leading to even more deficits, higher taxes, and more unproductive public spending. This vicious circle of unemployment, taxes, and public spending began in the 1980s and was worsened by the monetary policies of the 1990s, which led to high interest rates and an overvalued exchange rate. That undermined investment, made French production less competitive, increased unemployment, and led to recession. It also led to stagnant salaries and a massive increase in the public debt. If France had in the early 1990s pursued the same monetary policies as Britain, France's public debt would probably not be much higher than Britain's debt today.

I'm not saying this just to be controversial. I'm just trying to explain why we cannot afford to make the same mistakes as in the past and to underscore how urgent it is to use the single European currency to promote growth in the euro zone.

The reduction of the workweek has also had seriously negative effects. Whereas strong world economic growth in the late 1990s and early 2000s helped to restart economic growth and lower unemployment, the thirty-five-hour workweek put an end to rising salaries and halted our economic recovery. The large cuts in labor taxes necessary to compensate for the rising cost of labor contributed to the deficit. The development of State subsidies for jobs further contributed. These subsidies and the way in which they were linked to the cuts in labor taxes had the effect of holding down salaries, often to the level of the minimum wage. Why should a company raise salaries if it can get the State to pay instead? Why would it raise salaries if this would prohibit it from benefiting from cuts in taxes on labor?

The thirty-five-hour workweek not only contributed to the lack of salary growth. In many cases it also led to a reduction in overtime pay, which for many workers had been an indispensable supplement to their salary. What is the point of having more leisure time if you have less money to spend and if it's difficult to make ends meet?

Another negative effect of the thirty-five-hour workweek was the way it disrupted the organization of the public sector. It is impossible to overstate the degree to which its effects were disastrous, especially in hospitals. How could it have been otherwise when the law required a reduction in working hours but the maintenance of the same number of workers?

The thirty-five-hour workweek may well have created some jobs in the short term. I am certain, however, that it destroyed or prevented the creation of even more in the long term.

These negative effects were aggravated by the government's efforts to curb growing deficits by resorting to budgetary cuts. By blindly cutting spending every year to try to show the smallest possible deficit—often by using accounting tricks—the government frequently cut the most productive expenditures and reduced public investment proportionally. During this same period, current spending continued to race out of control. Much was wasted in efforts to delay inevitable costs and to spread out programs in order to meet annual targets. All this led projects to become more expensive than they should have been. When you put off spending on the Rhône seawalls and they end up giving way, the decision ends up costing a lot more than the minor savings made in the short term.

But the worst thing is that in France, the tax burden and the costs of social insurance fall mainly on labor. Labor bears the costs of just about everything. It pays for all the mistakes of our economic policies. With rising costs, labor becomes more expensive, jobs are destroyed, and costs rise even more.

The destructive effects of the overtaxation of labor have become suicidal in the context of a globalized economy in which French workers are in direct competition with workers from developing countries where living standards are much lower, social insurance does not exist, and currencies are usually undervalued. Every year, China and India add more than 20 million low-paid workers to the global workforce, lowering the global cost of labor.

While most of the world has been seeking to meet the challenges of globalization by working more and investing more, we have managed to undervalue work. Jobs have become increasingly insecure. We have more and more poor workers. Working conditions have deteriorated. The Nobel Prize–winning economist Edmund Phelps has noted that France is now the country in which workers are least satisfied with their jobs. We have so perverted our social model that instead of protecting people it has become the cause of insecurity and suffering. It's not hard to understand why so few other countries want to adopt it.

The revaluation of work is the key to the return of dynamism and growth. There are plenty of historical examples of countries emerging from the doldrums. In 1958, it took France only a few months to emerge from a political, moral, and financial crisis that seemed inextricable. But for that to happen, the vicious circle must become a virtuous one. The process of job destruction must be turned into a process of job creation. No one is going to do this for us. We must seek renewed growth from within. We must produce it with our energy, our imagination, and our work. The role of policy is to act as a an impetus for the energy, the imagination, and the work of the French. That's the goal I have set for myself.

GOVERNING DIFFERENTLY

We must first steer clear of false solutions that would actually make things worse rather than better. For example, France will not emerge from its current troubles by changing its Constitution yet again. In 1958, the institutions were part of the problem. The Fourth Republic was a weak and unstable regime, whose governments changed every two or three months in the wake of partisan political machinations. In establishing the Fifth Republic, General de Gaulle saved the Republic—which was on the verge of civil war and bankruptcy—for a second time. Since then, the Fifth Republic Constitution has enabled France to be governed under all kinds of circumstances. Perhaps one day it will be necessary to take the five-year presidential term, inaugurated in 2000, to its logical conclusion with a rebalancing of constitutional powers. But we need more time to reflect on that, and in any case it is certainly not by returning to the Fourth Republic under the guise of creating a Sixth Republic that we will fix France.

I am convinced that we must change our actions and our policies, but I am opposed to disrupting our institutions at a time when France most needs a clear, long-term orientation. The difficulties our country is struggling with are due not to institutional failure but to intellectual and moral failure. Focusing on institutional reform is a faulty solution that results from a faulty diagnosis of the problem.

This is why, unlike other candidates in the 2007 election, I did not go around proposing to put all sorts of electoral promises in the Constitution to give the impression that they would be kept. Public skepticism about politics will not be overcome by turning judges into the guarantors of politicians' sincerity.

It's not by walking away from politics that you can restore its capacity to inspire confidence. Confidence can be restored through sincerity, honesty, and promises kept. These are the values I believe in and the values that I want all the French to share. The Constitution has nothing to do with it. It's not meant to bail out those who don't have sufficient confidence in their values.

We don't have to change the Constitution, but we do have to govern differently. Parliament needs to play a greater role. The president of the Republic must govern. Competence must prevail over scheming, and the most important nominations must be subject to oversight by and approval from parliament. The government must be more collegial—it should be more than a collection of feudal administrative spokespersons who don't cooperate and who leave decision making in the hands of ministerial staffs and the technocracy. And to be more collegial, a government must be smaller. We need to have fewer but stronger ministers, who are better able to stand up to the bureaucracies and who can impose political compromises. Ministerial structures must be stabilized rather than changing continually as they do today. They must also be better adapted to the big challenges of the twenty-first century. These great challenges require us to place the focus of politics on life, intelligence, knowledge, globalization, sustainable development, social cohesion, integration, and national identity. As president I want to organize ministries around these priorities. I want to make possible a new type of government.

Having thought a lot about this on the basis of my own experiences, I am also convinced that the Finance Ministry should not be responsible for economic development. It should focus on the budget. The Finance Ministry should be the ministry of budgets—including national, social, and regional budgets. Within the government, the finance minister should play the role that the chief financial officer plays in the governing structure of a company.

But France's fiscal recovery will not result only from action by the Finance Ministry. The budget deficit is a consequence of public policy and of the malfunctioning of both the State and society. The fight against deficits is everybody's business. Of course we must go after the massive amount of tax evasion, which demoralizes those who honestly do their civic duty. But it is even more important to reform the State. We have to evaluate the effectiveness of public policies so that we can cut useless spending and create room to maneuver. We cannot hope to cut the number of tax officers and state accountants significantly if our fiscal policies make our tax code ever more complicated—it is already one of the most complicated in the world. We cannot reduce the number of State employees who oversee state aid to companies so long as the policy on State aid is incomprehensibly complicated. We cannot cut subsidies without evaluating those that meet their objectives and those that don't. These evaluations will be based on a logic of results rather than a logic of means. This will constitute a revolution in French mentalities which runs up against ferocious resistance from the fiefdoms that reject any self-assessment.

I intend to put this revolution at the heart of my plans to reform the French State because no one has the right to waste the money taken away from French workers, and because everyone must account for the effectiveness of what he or she is doing. Otherwise, the very legitimacy of taxes and national solidarity is called into question.

Of course, when governments have to deal with the difficulties created by a poorly performing society and economy, they struggle to deal with social needs, the costs of operations rise, and problems multiply. Thus we need to fight on both fronts—administrative reform along with economic and social reform—at the same time. We have to invest in reform, because reform can actually save a lot of money in the long run, even if it's costly in the short run.

The reform of higher education and research is vital. I believe it can no longer wait. But it can succeed only if we devote significant resources to it from the start. This is an investment from which the country—and ultimately even its fiscal position—can really benefit. Failing to make this investment would in the long run have incalculable costs and would compromise our future growth.

We have for too long believed that budget cuts would produce reforms by putting pressure on spending. But this strategy has failed. It has become the excuse for conservativism and weakness.

The truth is that in many cases it is investment in reform that will lead to savings and growth, not the other way around. As president I am betting on a policy of investment and reform to enable France to restart the economic growth which has been lacking, and which will enable it to overcome social divisions and to make the deficits disappear.

I am convinced that certain investments in education will in the long run lead to a reduction of deficits, not their growth. These include investments to reduce the number of pupils per class in disadvantaged neighborhoods, to provide for supervised study of homework for "4:00 p.m. orphans," to build top-quality boarding schools for talented students from poorer neighborhoods who cannot work at home, and to create "second-chance" schools (where young adults who did not do well in school can get further education) all across France.

I'm convinced that a Marshall Plan to provide job training for disadvantaged youth, as I have proposed, is an investment that would bring in more than it costs, because youth unemployment is such a waste.

The debate about public spending is often obscured by the fact that it is easier to measure spending than it is to measure the return on investment, because the cost of social difficulties is too diffuse to be easily measured or because it is simply deliberately ignored.

It is difficult to determine precisely the cost of the rising number of increasingly young children leaving school, just as it is difficult to measure the cost of pollution, stress, loneliness, traffic, worsening working conditions, or car accidents. But when you look at the studies that have been done, despite all the uncertainties with them, the enormity of the waste is striking.

Attacking these serious and urgent problems is like financing research or business creation: it's an investment, not a waste of money.

France is paying the price for not having invested enough in the past 25 years. I want to pursue a budgetary policy that respects the commonsense rule according to which only spending on investment can be financed by borrowing. But I want the decisions about what qualifies as investment spending to be based on sound economics, not on some bureaucratic accounting system.

I want the discussion with parliament over the national finance law to focus on whether a particular type of spending should count as investment from which returns are expected or as consumption that should be financed by current taxes because it will not produce any future revenue. This approach is consistent with the desire to control public spending and reduce the deficit. In fact it's a precondition for this, since it will enable us to spend more wisely.

There is no point in running around screaming hysterically about the debt. Real discipline means having the courage to cut wasteful spending and to make priorities.

To get France going again, to pay back the debt, and to pay our future retirees, we need to work more, and not simply make the French people tighten their belts. We don't need a policy of sacrifice; we need a policy of effort.

Work creates work. Work creates wealth, well-being, the feeling of being socially useful, and self-esteem. The only way we'll surmount France's crisis is by putting more emphasis on work.

I am convinced that putting more emphasis on work is the key to our future. I want to make it the top priority and the key to all our public policies. This is not merely a technical matter but a moral one. It is a question of values—a human and social issue, a question of civilization itself.

The policy I am proposing is more than anything else a moral choice. The core of my policy is compensation for effort, merit, and risk. It is about providing the means necessary for talent to develop—equality of opportunity means the possibility of a fresh start for those who have failed.

The mechanisms for this policy are salaries; changes in tax, budgetary, and monetary policy; and also social insurance, schools, and training. But the place to start is with a new pact of confidence between the working public and the nation.

BUYING POWER

At the end of the 1970s, after a long period of inflation during which buying power rose much faster than national income—to the point that the return on capital and corporate self-financing had become insufficient—Prime Minister Raymond Barre began a courageous policy of de-linking salary increases from prices. The goal of this policy was to halt the price-salary spiral and to give companies a greater share of profits. Other than the debacle of Mitterrand's economic experiment of 1981 and 1982 and the economic recovery from 1997 to 2001, this policy of austerity for salaries has lasted to this day, well beyond its necessity. Whereas the entire economy is indexed for inflation and all prices are adjusted for the adoption of the euro, only non–minimum-wage salaries have remained unindexed, with a chain reation of effects for the rest of the economy. Over the past twenty-five years, rising prices have lowered the living standards of workers, notwithstanding the apparent rise in salaries. This erosion of buying power has been hidden in the statistics by the rise in the level of qualifications and by the difficulty of measuring the evolution of the cost of living based on consumer price indexes when not all prices vary in the same way, with some falling whereas others rise. The French are well aware of this gap between the statistics that claim that buying power is rising and the experience of daily life that proves the opposite.

Today the rules governing indexation are affected by all sorts of exceptions. I propose reestablishing contractual freedom and allowing for these rules to be freely negotiated in each field and in each company. It makes no sense for government bonds or rent to be indexed while non–minimum-wage salaries are not.

But the goal of the policy I'm proposing is not to keep buying power stable; it's to increase it. The goal is for the children of tomorrow to have a better life than their parents. The goal is not stagnation but progress. It's not the minimum but the maximum.

For salaries to rise we have to act to counter the deflationary effects of competition from countries where salaries and currencies are undervalued. This is why the overvaluation of the euro must be fought. The overvaluation of the euro undermines European workers. Companies have to limit salaries to compensate for the overvalued currency. Currencies are supposed to absorb shocks, not amplify them. The goal of monetary policy should be to lessen the effects of competitive depreciations of our competitors rather than to aggravate them. Even while affirming that the right monetary policy will in no way allow France to get around the need to reform and work harder, I cannot accept that the efforts of French workers should be undermined by the overvaluation of the currency. I cannot accept that the euro should contribute to the destruction of our industry when it should be supporting it. All you have to do is look at Airbus to see how bad monetary policy can destroy the good work of several generations to build such a magnificent industrial base. One of the priorities of my European policy as president will be to change the direction of policy regarding the euro.

A more active policy toward the euro alone will not correct the monetary distortions that so seriously affect our interactions with the world economy. We will also have to negotiate a readjustment of exchange rates with countries that deliberately practice monetary dumping by selling their products cheaply thanks to an undervalued currency. Europe should have stood by the United States in trying to negotiate the revaluation of the Chinese currency. It will be essential for Europe to take part in this battle in the future. For this it will need not only the will to act but the means, by adopting a policy of "community preference," which could be leveraged in the negotiations, as the Americans leverage their own.

Beyond monetary policy, globalization requires us to think about our tax policy. We have got to reduce taxes, if only to bring our rates down to the European average. We can do this by cutting unnecessary spending and ceasing to waste the public's money. We can do it by cutting social spending through a policy of economic growth and employment.

The reform of our tax system can contribute significantly to this. It would give people the incentive to produce and would provide the competitiveness our economy needs in the face of increasingly tough international competition. I've got to be frank: in today's world, if we keep overtaxing work and capital, both will go elsewhere. The only things we can tax without fearing labor or capital flight are pollution and consumption.

In a world of globalization, to counter the deflationary effect of low-cost competition on salaries and to prevent jobs from being exported, the only option is to lower taxes on workers and investors, and to increase them on polluters and consumers. I feel a duty to speak this truth to the French: that in the world economy it is suicidal to overvalue your currency and overtax work and capital. This contributes to long-term unemployment, a decrease in buying power, and social exclusion. It prevents our industry from being able to compete with that of emerging countries. It leads to deficits and starves the State and public utilities of resources.

Given the crisis in France, we have to know what we want. I propose that exports from countries that do not respect environmental rules be taxed according to how much they pollute. One could also imagine a tax on the carbon content of imports. Even more imaginatively, we could put in place an antipollution tax system which would impose taxes on imports and national production but which would be deductible for exports. At the same time, production and consumption of ecological products would be encouraged by a lowering of the value-added tax (VAT). I intend to propose to our European partners that we begin discussing this issue of fiscal environmentalism.

Having thought a lot about it, and having studied the experience of Denmark and more recently of Germany, I am convinced that we must experiment with the transfer of some social charges to the VAT, since such a system allows you to exempt exports from the financing of social insurance and to tax imports instead. This would be the opposite of the current system for financing social insurance, which, by taxing work, burdens our exports, increases the price of French goods compared with imports, depreciates the reward for work, and undermines employment. It is time to start experimenting with this sort of thing. We have waited too long. Our German friends are at this very moment showing how promising this approach can be.

Financing social insurance with a tax on sales, sometimes rather bizarrely called the "social VAT," has many advantages in the context of the global economy. It is a way to fight against exporting employment, to create jobs, and to boost buying power. By raising the price of imported products, it lowers the incentive for consumers to buy cheap imports from low-salary countries rather than products made in France by more expensive labor. The consumer, who is all the more driven to find the cheapest goods because his buying power is weak, is thus given an incentive to participate in the destruction of his own job and social insurance. By taxing imports and exempting exports, the transfer of certain social charges to the VAT has the same effect as a devaluation, which can attenuate the effects of the overvalued euro and boost competitiveness. By lowering the costs imposed on labor, it would make it possible to dampen the fluctuation of employment when growth slows. Cutting the cost of work would most benefit those companies who hire the most workers, without thereby penalizing high value-added companies vis-à-vis their foreign competition. As the VAT covers a wider range of transactions than social charges, around one-third of the decrease in charges could be reallocated toward buying power. This, in turn, would contribute to a rise in salaries.

Would that lead to inflation? I do not raise this issue lightly, as prices in France are already very high and that is why we need to experiment. The cost of imports would rise, and this is one of the desired goals; but it is doubtful that other prices would rise, because the cut in social charges would compensate for the rise in the VAT. On average the price of goods made in France would be likely to remain the same, given competitive pressures. The German experience with this is encouraging.

But I understand that the risk of a rise in prices that would undercut buying power is a cause of concern to workers and retirees, who are so used to being deceived and having to absorb costs. To avoid any such concern, I propose that with the agreement of workers and industry, every step toward the social VAT would be followed by a conference at the end of each of the following two years. This conference would focus on salaries with the mission of examining the evolution of buying power of workers and retirees and to compensate for the eventual losses that might be attributable to the social VAT. Thus, workers and retirees would be assured that they would not yet again be the victims of machinations whose goal is to make them pay for others. For I am convinced that nothing can succeed without the confidence of workers. And this confidence today is shaken.

Will the social VAT be unfair? I don't think so. It will in any case be no more unfair than the social charges that companies withhold from salaries. Will the rise in the VAT promote moonlighting? I don't know why it would lead to more tax cheating than the social charges that it would be substituting for.

If the experiment proves successful, the transfer of the tax burden onto the VAT should be done gradually, by increasing exemptions on taxes. I invite our other European partners to experiment in this area where Germany has preceded us. Along with a Europe-wide environmental tax on imports, this would be a good way of putting in place a sort of community preference. It would allow us to consolidate the European social model without giving way to the temptation of protectionism. France invented the VAT, which was then gradually adopted by all European countries. My objective is for us to do the same one day for the social VAT. This will inevitably lead us to reconsider the way we provide social insurance. It will force us to begin dismantling the system of differentiated levels of social insurance and will make all the French equal in terms of health insurance and retirement. Within the social insurance system that covers all of the French, it will also take better account of labor mobility; multiple employment; and the different phases of work, unemployment, and training in a given individual's life. The reform of our special retirement schemes that I'm calling for is consistent with this approach, as is the complete overhaul—which this reform will enable us to finance—of the criteria for hardship and the setting of retirement income for farmers, shopkeepers, artisans, and widows.

The social VAT would be not merely a new way of financing social insurance but a lever for change, an incentive to change our economic behavior and our mode of social organization to meet the challenges of globalization. This is why I want to give it a try. Beyond the particular case of the social VAT, what's most important is to tax work and production as little as possible. We must tax the wealth that is produced, not the production of wealth. When you discourage the creation of wealth, you have less to redistribute.

One of the keys for renewed French success in the global economy is a major reform of taxation, which will doubtless have to take place over time but which is absolutely necessary. We have already waited too long to do this, and we are now paying a price in terms of jobs, growth, and buying power. This reform is the necessary condition to make protection and competitiveness compatible. To back away from competitiveness would impoverish us. To back away from protection would be a burden for our economy because the fear of risk would become too great. Our goal must be for all tax cuts primarily to promote employment. If we really believe that only by reemphasizing work can France recover, prosper, find renewed cohesion, and overcome its deficits, then we must do everything we can to boost productivity and reduce the difference between the cost of labor to employers and workers' buying power. This is the only way to make work more competitive internationally while increasing workers' living standards.

The other option—trying to increase competitiveness while diminishing workers' living standards—is a dead end. You can't base French workers' living standards on those of Chinese or Indian workers. Revaluing work means moving away from a society in which government assistance pays better than work, and in which a very small number of people enjoy the fruit of the efforts of all. From this point of view, the question of buying power is central. Poor workers and the fear of falling behind that grips those in what we call the middle class are constant reminders of this.

As president, I want to make retirement a choice rather than an obligation. I will allow the retiree who wants to work to do so freely, with no restrictions.

I will provide a grant for all young people starting at eighteen years of age, as long as they're getting job training and following it diligently. I will create zero-interest loans that can be reimbursed from future earnings so that they can pay for training or job creation. I will ensure that everything be done to allow young people to study and to do a paid job free of taxation. This way each student will be able to attain financial self-sufficiency and will become completely responsible for him- or herself. I will engage workers and companies in a discussion of overhauling our welfare system to get us beyond the poverty and inactivity traps that exist today. We'll have to do away with the segmentation, the complexity, the lack of transparency, and the incoherence of the current system, which in many cases contributes to social exclusion instead of helping to overcome it.

I will ensure henceforth that income from welfare can in no case be greater than income from work and that welfare will not be paid unless the person receiving it, if possible, does something in return. This could be some public service activity that would preserve individuals' feeling of social usefulness.

In the same spirit, we should stop subsidizing an increasing number of salaries in a way that simply substitutes for greater pay and contributes to flattening pay levels around that of the minimum wage. The original purpose of these subsidies—to give an incentive to create new jobs—must be restored. There are today eight million beneficiaries of this subsidy! That means we're failing to choose and that yet again we're just throwing money away.

The threshold effects created by lowering labor taxes have to be eliminated. They lead employers to freeze salaries in order to get tax exemptions. I would prefer to see a system of exemptions negotiated between companies and workers that would gradually replace that of thresholds.

Profit-sharing and stock ownership must be made possible for workers who have contributed to the success of the company, if the workers want this. Stock options or free shares, which constitute another type of motivation and compensation, should not be restricted to just a few. As all the workers contribute to the success of the company, they should all be able to share in these benefits as well.

We also need to cut the succession tax on the fruits of a lifetime of work that has already been taxed all along. This is simply a recognition of the value of effort, and it would encourage people to work more in order to leave more for their children.

Another good way to encourage work and success would be to establish the principle that no one should pay more than 50 percent of his or her income in direct taxes. This would put an end to a fiscal policy of confiscation that chases out of France the gifted, the creators, and the entrepreneurs. It also drives out capital, savings investments, and the working spirit.

My proposal is to get beyond the logic of sharing and scarcity and instead to embrace creation and growth. Those who want to work more should be able to do so—they should even be encouraged to do so instead of being limited to thirty-five hours per week. This is why I think overtime should be paid a bonus of at least 25 percent and exempted from social charges and taxes. Someone who wants to work only thirty-five hours per week because he or she wants to give priority to life outside work will be free to do so. But someone who decides there's no point in having more vacation time if he doesn't have the means to pay for a family vacation, or who finds himself at a time in life when he needs more money, will be able to work more and earn more.

This will be a great revolution in behavior. It will provide freedom of choice and compensation for effort. It will mean setting the legal limit for work as a minimum and no longer as a maximum. For those with low salaries it will mean the possibility to significantly increase their buying power. This behavioral revolution will bring about a social and an economic revolution. The increase in work for some will not take away jobs from others. On the contrary, it will mean more buying power and more consumption, more opportunities for companies and more economic activity. It will mean more jobs and more revenue for the State.

AGAINST LAISSEZ-FAIRE

The anxiety among French workers today stems not only from the devaluation of work and the competition from globalization but also from the excesses of financial capitalism where speculation can become more important than production, where the financial predator can become more important than the entrepreneur, and where the demand for profits becomes so great that it can no longer be satisfied without sacrificing the long term in the name of the short term. We cannot fight against this destructive logic on the national level alone. This is why I have proposed that Europe join the fight. I think that issues like tax evasion, golden parachutes, and uneconomic and destructive hostile takeovers should be dealt with at the level of the euro zone. I intend to demand that companies that do business in France be more transparent and that shareholder groups have more of a say over executive salaries.

The concept of laissez-faire is alien to me. I believe in the creative power of markets but I do not believe that they're always right.

I want risk to be shared. Guarantees should be given to those whom the free market prevents from taking risks because they don't have the right connections or enough resources.

On the model of how we saved Alstom when I was finance minister, I want the State to be able to step in temporarily to buy shares in strategic companies to help them get through a difficult period or to enable them to escape from predators whose main objectives would be to empty them of substance or get access to their industrial secrets.

To revitalize family capitalism in the face of stock-market capitalism, the tax system should support family-owned companies instead of penalizing them, as it does today. Investment in small- or medium-sized companies should be encouraged by giving a choice to the taxpayer—within a limit of 50,000 euros—of either paying a wealth tax or investing in such companies.

The tax system should be an incentive for and not a brake on investment and job creation. This will be the case if we stop taxing the creation of wealth and start taxing created wealth. But it will be even more the case if we raise the research tax credit to 100 percent to promote investment in innovation—and even more if we design taxes so that companies that cut jobs and disinvest pay the most while companies that add jobs and invest pay the least. Such an incentive has the advantage of not penalizing companies that are struggling, which, by definition, don't have any profits and are not paying any taxes. It seems to me perfectly reasonable for a company that leaves the country to have to pay back the State subsidies that it has received, just as it seems reasonable to me that a profitable company not be allowed to benefit from a cut in taxes if it does not raise salaries. We cannot expect to make capitalism more moral if we accept this sort of behavior.

In this activist policy I'm proposing, State-owned companies must play an active role. They have an even greater responsibility to the nation than others, and they must fulfill it. It is not acceptable for public ownership to be managed like any old stock portfolio focused on short-term profits with little regard for the strategic interests of the stockholding State.

Public procurement must also be more systematically put to use to develop small- and medium-sized companies, as in the United States. This is said to be against WTO rules, but the United States has an exemption. Why should Europe not benefit in the same way?

I will not resign myself to allowing the financial markets to decide everything for everyone. In the current state of malaise, there is a feeling of alienation. Many, especially the young, are suffering because they feel obliged to leave the lands of their childhoods, where their families have sometimes lived for generations, not because they want to see the world but because the local economy has collapsed. They all gather in the suburbs of the big cities where they struggle to find jobs and homes as their original towns and villages die out after having been homes to people for millennia. This is absurd. France's interest is to reindustrialize declining employment zones rather than to resign ourselves to this exorbitantly expensive economic and human waste. The State would do better to invest in order to create new activity in these areas, which have a strong working tradition and an industrial culture, rather than to waste billions of euros in public money for early retirement.

France and Europe need a new industrial policy. Put another way, we need a real policy for the entire system of production, its capacity for innovation, its specialization, its regional distribution, its protection from predators, and its financing. We need a forward-looking, collective economic strategy. The response to globalization requires a comprehensive, coherent, determined policy for production and employment that provides opportunity to all individuals and regions.

A NEW ECONOMIC POLICY

This policy of breaking with the ideas and practices of the past—with the conventional wisdom that has let us down for 25 years and with the habits that no longer meet current needs—cannot be put into place without the agreement and participation of the entire country.

I want to begin a national dialogue about all of this. I believe in social democracy. But social democracy must be as irreproachable as political democracy, and as in any democracy worthy of the name discussion must prevail over confrontation. I want to widen the use of the "social alarm" system that has worked well in the Paris Metro. But I also want to stress that for me, in a social democracy worthy of the name, the sacred right to strike must not mean that public sector workers are able to take the country hostage. This is why I plan to ask Parliament to support a law that would provide for minimum public services in case of a strike in the state sector.

In a social democracy worthy of the name, union representation must be real, not theoretical. We cannot continue basing our social democracy on a notion of union representation that is based on a patriotic attitude from World War II. I will thus ask the government to prepare a law that would lift the restrictions on who can run in the first round of votes for union representatives and would set new criteria for representation based on elections.

In a social democracy worthy of the name, a minority may not impose its will on the majority. I therefore want to require, after eight days of a strike, votes based on a secret ballot that would have no effect on the right to strike but would allow all to know where they stand and which would prevent a minority from speaking for the majority.

This renovation of social dialogue will open a new era in the social relationships that have always been difficult in France.

I want cooperation and negotiation among companies and workers to precede parliamentary discussion. I want us to rethink the means and instruments of this social dialogue.

I am convinced that no policy of change and no major reform will be possible without the involvement of all, just as during the Thirty Glorious Years when our economy succeeded.

I intend to propose that French business and labor re-create a place where the State, unions, business leaders, experts, scholars, and intellectuals can again not only meet but participate in real dialogue, reflect together, anticipate change together, imagine future society together, and work together on the big, collective choices that must be made for the long term. We are living in a period of change similar to that which followed World War II, when we had to rebuild everything to be able to meet challenges of a very new nature, when we had to reinvent everything because our old methods and our old ways of thinking had been passed by.

CHAPTER 11

Rethinking Foreign Policy

ADAPTING OUR INTERNATIONAL MESSAGE

If there is one area about which I really think we need a serious and open debate it's foreign policy. This is in no way a criticism of the diplomacy of former president Jacques Chirac. Jacques Chirac devoted time, energy, talent, and experience to foreign policy. His stance on the Iraq crisis spared France some serious unpleasantness, and it should have been better accepted by our American friends. But this shouldn't stop us from reflecting frankly on the main orientations of our national message.

Let me repeat first that it should no longer depend on the will of a sole individual, even the president of the Republic. The very notion of *domaine réservé* is meaningless in a democracy that wants to be an example to others. Our foreign policy would also be more coherent if it were debated and won support that went beyond whatever majority is in place at any given moment. It would thus be more effective. Finally, while still believing in France's universal message, I fear that by wanting to be everywhere, on every subject, all the time, we run the risk of exhausting ourselves and forgetting where our fundamental national interests lie. Here, too, we need to know how to make choices—first how to think about them, then how to explain them. While France's message is relevant for the entire world, our strategic interests are different in different geographical areas. There are countries where our presence is vital to our future.

I therefore think it is necessary for us to rethink our traditional economic relations and to reorient them toward areas of high growth. China, India, Brazil, and Southeast Asia should be priorities. If we limit most of our foreign trade to our geographical neighbors, we can expand our exports only as fast as their economies grow. But these economies are not growing quickly. Thus we're depriving ourselves of numerous opportunities. Our foreign policy needs to adapt to new international realities more quickly. The geography of global economic growth has been turned upside down over the past ten years. Our diplomatic network should be inspired by this. It should question itself and adapt to this new context. The same goes for culture. I am not sure that we need foreign economic agencies and several consulates and branches of the Alliance Française in every EU country. On the other hand, we do need to deploy these agencies and spread the French language and culture in places such as India, China, and Brazil.

EUROPE

Europe is a project of peace and civilization. French culture would not survive the death of European culture. If Europe fell apart, the European conception of freedom of thought and human dignity would also disappear because no European nation would be strong enough to make its voice heard or to resist global cultural harmonization.

Europe cannot simply be a place you come from. To count in the world, Europe must be ambitious. André Malraux was right to say that "Europe must be ambitious or it will die." The EU's found-

ing fathers—Jean Monnet, Robert Schuman, Alcide de Gaspari—
had this ambition. So did Winston Churchill, Charles de Gaulle,
and Konrad Adenauer. And so did Georges Pompidou, Valéry Giscard d'Estaing, François Mitterrand, Jacques Chirac, Willy Brandt,
Helmut Schmidt, Helmut Kohl, and Jacques Delors. From the European Coal and Steel Community to the single currency, Europe
was built on the ambitions of its people and governments.

These people and governments wanted a Europe that could
act, not a passive Europe. They wanted a Europe that would multiply their power. They wanted a Europe that could protect them,
not a Trojan horse. They wanted a strong Europe that could defend
their interests and values in the world economy, not a victim of globalization. They wanted a democratic Europe that would respect
the national identities and sovereignty of its people, not a bureaucratic Europe that flattens everything under its regulations, that
prevents any industrial policy in the name of a dogmatic vision of
competition, and that bans macroeconomic policy. They wanted a
single currency that works for Europe's competitiveness, growth,
and employment, not a Europe stifled by an overvalued currency.
They wanted the currency to work for the economy, not the other
way around. They wanted a Europe with personality, an identity,
and borders. They wanted a Europe in which they could recognize themselves—a European Europe. They didn't want a Europe
with no fixed borders, expanding indefinitely, diluting its institutions, policies, and will in an ever-wider, heterogeneous, and loose
grouping.

The enlargement of EU has weakened the common will and
placed an insurmountable obstacle before political integration. Turkey's entry would kill the very idea of European integration. Turkey's entry would turn Europe into a free-trade zone with a competition policy. It would permanently bury the goal of EU as a global
power, of common policies, and of European democracy. It would be
a fatal blow to the very notion of European identity.

Part of our current identity crisis results from a Europe that
has come to symbolize weakness, with an overvalued euro, a free-

market ideology, and a dogmatic commitment to competition. It results from endless enlargement, a "race to the bottom" in fiscal and social policy, and the feeling of many citizens that Europe is being built without them or even against them.

I have always supported Europe. I supported the single currency and the European Constitution. But I do not believe that the rejection of the EU Constitution was a rejection of the outside world. I don't think it meant the French were turning inward or getting cold feet. It was the scream of a France that has had enough of others deciding for it and of hearing leaders claim there is nothing they can do. By using Europe as a pretext for all our unkept promises, we turned the "no" vote into a vote against unkept promises. In fact, it should have been the other way around. Saying "yes" to Europe should have been seen as a way to ensure that promises are kept.

We can no longer continue to construct Europe in the same way we did before the "no" vote. European elites can no longer keep European policy away from the general public on the grounds that it is too important to be left to ordinary citizens. We can't just go on telling the France that voted "no" that it was wrong, that the EU is too complicated, or that it didn't understand the issue. Europe cannot be made without or against its citizens. The France that voted "no" was expressing pain and anxiety which go well beyond the question of Europe, but in which Europe's drifting off course has played a major role. By judging and condemning the France that voted "no" instead of understanding it and taking it seriously, we risk pushing it to rebel even further.

Reconciling the France that voted "yes" with the France that voted "no" is my main focus. It is a necessary condition for reconciling France with Europe and for getting Europe to help solve the crisis rather than make it worse. This is my priority. Europe's weakness is not inevitable. All Europeans, and not just the French, critically need a Europe that fulfills its responsibility and protects its citizens.

The Europe of the Founding Fathers believed in itself. It believed in its values and in its purpose. It put its coal and steel together, organized a common agricultural policy, established the principle of community preference, and made Airbus possible. Now the future is not in steel but in areas like energy, the environment, biotechnology, and space. Yet the same kinds of initiatives are necessary. Europe can't put its energy future in the hands of a few traders in London, New York, or Singapore. It cannot allow the sort of giant profits and investment shortfalls that will prove very costly in the future.

Industrial Europe is not one that obstinately blocks the creation of European champions and that prefers for companies to be bought up by American or Indian companies rather than European ones. Industrial Europe should help companies become more competitive and improve their products.

A Europe that is not afraid of taking responsibility is not one that buys into the religion of free markets. It is one that gives a priority to European products.

A Europe that has self-confidence is one that adopts the principle of subsidiarity (in which decisions are taken at the closest level possible to the people), which has an economic government, which limits its enlargement, and which abandons the unanimity rule.

This Europe is one in which no one can force a member state to join a policy it doesn't want to. But it's also one in which no state can prevent the others from acting. It is a Europe of those who want to again express a common will, of those who want to act and not submit. This Europe is possible. And it is necessary. That's the Europe I want to promote. A Europe that will reconcile the French with the EU.

Before we remake Europe politically, we have to remake it economically and socially. In the current situation, if we want to realize the European dream we have to redefine the principles and rules of the economic and monetary union by giving them this human and social dimension that is currently lacking.

No one will manage to restore faith in Europe if it is perceived as a source of poverty and not prosperity, if it is seen as a force for economic and social regression and not progress, and if it cannot win over hearts as well as minds.

The French said no to the European Constitution because they had the impression that Europe didn't protect them anymore and that it made them victims rather than masters of globalization. The best response is to remake a Europe of common policies rather than to continue to make a nonpolitical Europe. I intend to propose to our European partners that we create a genuine economic government of the euro zone. I will propose that we reopen the dossier of the Common Agricultural Policy (CAP) with the objective of guaranteeing European food security (which is today far from the case) as well as sanitary security, respect for the environment, and a decent income for farmers. Rethinking the CAP is a necessity. Getting rid of it would be catastrophic. That would make agriculture subject to speculation.

I will propose to our partners that we adopt a common energy policy so that together we can confront gas and oil scarcity. Europe should not stand in the way of more powerful electric companies. It should look after the interconnections and investments in infrastructure and in production, which is falling behind needs.

In the area of natural gas, Europe should not question the right of Suez and Gaz de France to merge, but rather create a centralized European gas company that would be able to stand up to giants like Gasprom.

I will propose an environmental Europe, not to burden European industry with niggling regulations but to invest massively in its own technologies and to tax the carbon content of imports from countries that don't respect environmental norms.

We have to remake economic Europe. But to do that we have to get out of the current impasse. Because Europe has stalled. It has stalled institutionally because there are now too many member states for the unanimity rule to function and because national interests are more and more diverse. It has stalled because people in many countries are not on board, and no state is strong enough to lead the others.

Getting European institutions working again is for me an absolute priority if we want to prevent the EU from turning into a simple free-trade area, where the world's speculators and predators would do battle.

ABOUT THE AMERICANS

I would like to put special emphasis on our relations with the United States. Our situation is unique. The United States is a country that some of France's elites claim to detest, or at least criticize regularly and in a stereotypical way. This is rather strange for a number of reasons. The United States is a country that France has never been to war against, and there aren't so many of those.

I stand by France's friendship with the United States, I'm proud of it, and I have no intention of apologizing for feeling an affinity with the greatest democracy in the world. It goes without saying that this friendship does not prevent either side from making its own assessments and taking independent action. And since this goes without saying, I don't feel it necessary to run around repeating it every chance I get. It often seems to me that the more you assert your independence with words, the less independent you are in reality.

France and America are bound together by unbreakable historical links. People often forget that the Revolutionary War, which led to the creation of the United States, was long and difficult and that its outcome was uncertain for some time. But France was right there at America's side for the decisive battle of Yorktown in 1781, and it was a young Frenchman, Lafayette, who led the final attack on the English camp. Without French support, history might well have followed a different path, one that would have been less favorable for the development of human freedom.

In the twentieth century, it was America's turn to protect France's freedom on several occasions. In 1917 and again in 1944 hundreds of thousands of young Americans crossed the Atlantic to pull Europe back from the verge of collective suicide. The French cannot forget that it was the Americans who liberated them from Nazi barbarity and who put an end to the bloodletting that this regime inflicted on the whole of Europe. For the forty years that followed the war, during which another kind of totalitarianism—communism—engulfed Eastern Europe, it was the military alliance with the United States that enabled France and Western Europe to preserve their freedom. After centuries of hatred, after the Holocaust, European nations embarked upon one of the most ambitious projects of their common history: to create a zone of peace, unity, and solidarity. The United States was always at the forefront of this project, supporting it politically and financing it with the Marshall Plan, which protected Europe from Communist imperialism.

I am particularly sensitive to this gift of liberty in several ways: as a Frenchman, as a political leader who has always worked to promote freedom, and finally as a son who wants to honor his father, who settled in France in 1948 after fleeing Communist Hungary.

France and America, then, stood side by side to defeat the two deadliest forms of totalitarianism in world history. And now at the start of the twenty-first century, the United States and France again stand together in the same camp against a serious threat to global freedom. It was the United States that was attacked by Islamist terrorists on September 11, 2001, but it could just as easily have been France. Indeed, many French citizens died that day in the Twin Towers. Terrorists do not distinguish among free societies. They want to destroy or subjugate them all, without distinction.

Every time that terrorism strikes—whether in New York, Madrid, Beslan, Tel Aviv, Casablanca, Amman, or London—it is freedom that is the target. Facing such a threat, free countries have no choice but to pool their forces and work together.

America came to aid and defend us twice in our recent history. We share a system of very similar democratic values with the Americans. And our children dream of learning about the American way of life and the things that Americans like doing. In addition, the United States is the leading economic, monetary, and military power in the world. We share use of the same ocean as America. You don't have to be a grand strategist to understand that our interest is to have the best possible relations with this country.

Everything should lead France and the United States to understand each other and help each other. But that's not the way it has been. Our relations have been cool, if not to say cold. I'm the first to recognize that the Americans are partly responsible for this. They tend to think that they are on the side of good and thus that everyone else is on the side of evil. They lack curiosity about and appetite for a world that, for many of them, stops at their own borders. They are certain that they are always the best. All of this can certainly be irritating.

But where our strategic interests are concerned, systematically opposing the United States is a double mistake. It's a mistake first of all because ignoring or criticizing your friends is bad strategy. And the Americans have been, are, and will remain our friends and allies. It's also a mistake because you can more freely express disagreements if you do so without questioning fundamental links. Thus on Iraq, our disagreements were legitimate, but they would have had more impact had they not been coupled with the threat of using our veto. France is strong enough to refrain from passionate, allergic, or excessive reactions. I believe we need to get along with the United States. I think we'll be all the better placed to express our disagreements—and they exist—if we can find a way to pacify and clarify our relations from top to bottom. I think we need to avoid confusing durable friendship with a people with whatever differences we may have with a particular government at any given moment. I am not overly fascinated by the American model. But if I had to choose, I feel closer to American society than to a lot of others around the world.

GIVING PRIORITY TO AFRICA

I am convinced that we need to consider Africa as a priority area. Geography inextricably links Europe to the African continent. At our doorstep, 900 million Africans represent the youth of the world. Four hundred fifty million of them are under seventeen years old. Their poverty and lack of a future are their problem today. Tomorrow these problems will be ours. Europe cannot remain a stable continent if it does not have the wisdom to help with Africa's development. This is not just a moral question but a vital challenge for Europe as a whole. No European country will be able to stand up to this challenge if Africans continue to believe that their economic salvation lies in Europe.

Africa is thus a priority, and not just francophone Africa, but all of Africa. While we're making a priority of it, we have to re-think the way we make Africa policy. If we want to show respect for Africans we need first and foremost to tell them the truth, speak frankly to them, and treat them as serious interlocutors. We've especially got to stop excusing them from all responsibility for the underdevelopment of their continent. Blaming Africa's failure only on the consequences of colonialism is contrary to reality. To do so is to refuse the sort of pragmatic diagnosis of the problem necessary for Africa's recovery.

We need to speak frankly with Africans about immigration. An agreement is possible. We cannot welcome all of Africa's youth into Europe, and they don't want to be dispossessed of their elites. Having us decide on our immigration and them decide on their emigration is all the more possible in that our interests are more convergent than many think.

We have to put a stop to the policy of relying on these famous "networks" that claim to love Africa but are really just exploiting its wealth and taking advantage of its weaknesses. These networks are corrupt, corrupting, and fake. They rely on corrupt practices that they have reveled in over the years. This is an odious image of Africa and France.

Between friends, allies, and neighbors, you don't need pseudo-official relations, because frank and friendly official relations can handle all subjects adequately. I would add that friendship with Africa must first of all be friendship between people who communicate by means of democratic institutions. Not everything can be based on personal relations between heads of state. These relations are precarious because they last only the length of a presidential term. They can become ambiguous. This is not the way we're going to develop the new African policy we so badly need. In the same spirit, we really have to favor democratic states over those that are not democracies. From this point of view, Mali and Benin are examples of countries we should strengthen, support, and help. It's time to understand that there is not one "Africa" but many "Africas." Favoring those whose democratic values are most similar to ours is not merely a possibility but a duty.

With this in mind, I would like to see the start of a debate about our military presence on the African continent. So long as it's a factor for peace, prevents genocidal clashes, and relaxes tensions, it is compelling. But let's be careful not to confuse things. The French army cannot get involved in internal African power struggles. Its job is not to stabilize regimes, back leaders solely because they might be pro-French, or favor one set of assumptions over another regarding a succession issue. The situation in Côte d'Ivoire should be a lesson to us. In this country, long considered to be the "Switzerland" of Africa, we run the risk of having bad relations with everyone. We're failing to reestablish order yet creating risks for our soldiers. And none of that prevented the indefinite postponement of the elections. I'm not in any way talking about abandoning Africa by depriving it of any French or international military presence. But we have to better codify this presence and ensure more transparency in its use. And we must never again hesitate to refuse to engage our military forces when democratic conditions are not fulfilled.

We must also urgently deal with the situation in Sudan's Darfur region, where thousands of poor people have been displaced and threatened. It must be made clear to those who refuse to apply U.N. Security Council resolutions that they will one day have to answer to the International Criminal Court for the fate of three million displaced persons.

The horror in Darfur takes place on a daily basis. The degree of violence and cruelty is reminiscent of the worst tragedies the African continent has seen in recent decades. For many of our citizens, Darfur is far away, but for me it's close. The men and women of Darfur may have different skin color from a majority of the French and Americans, but a different skin color does not justify abandoning people to an ignoble tragedy. How can great democracies such as the United States and France fail to be interested in Darfur's situation? Sudan must face up to its responsibilities. It is urgent to act so that Darfur does not remain a shameful page in our own history. Our indifference, our blindness, our lack of courage—or perhaps a combination of these things—must not lead us to avert our gaze from the first crime against humanity committed in the twenty-first century. The twentieth century saw many crimes against humanity. In Darfur, a crime against humanity is happening right now.

REALPOLITIK AND HUMAN RIGHTS

At the risk of seeming naive in the eyes of cynics, I believe in the necessity of preserving, incarnating, and defending our values in the international debate. Put another way, I do not belong to the realpolitik school, which says that you should sacrifice your principles on the altar of greater economic interests. At the top of the list of values to be preserved is respect for human rights. This is not in my view a mere detail. It's the foundation of the very notion of international community. A martyr is a martyr whatever the color of his skin and whatever his nationality. You cannot equate economic interests and respect for universal values. I don't mean we should seek to impose a model on others, preach, or hold ourselves up as the guarantors of good in an evil world. And I certainly don't want to contribute to a "clash of civilizations." I just think we need to be faithful to the democratic principles that require us to be frank.

I remember that during the cold war, people pretended to believe that the people of Central and Eastern Europe didn't have the same aspirations to freedom that we do. The Russians were destined to live in a dictatorship because after all they had never known anything else. It was supposed to be in their mentality. I don't believe in the cultural relativism of human rights, freedom, and democracy. I think these are universal values and that all people aspire to them.

Asking the Chinese about the fate of their political prisoners is not failing to show respect for the empire that China is. China is becoming successful enough that it doesn't need to take offense when the world asks it to explain its democratic shortcomings. You can admire a civilization, celebrate its recent and remarkable successes, and build a solid relationship based on deep friendship all while being clear and demanding in areas where silence is unjustified. Not speaking out is being complicit. And one of the great things about globalization is that it enables everyone to make good use of the same information. Today we all know everything about everything, at almost exactly the same time. Silence has become all the more odious.

What I say about China I could say about Russia as well. We must be sensitive about Russian national feelings, which have been so greatly tested over the past fifteen years. But we cannot and must not remain silent about the Chechen tragedy, the illegitimate Russian interventions in Belarus, the inexcusable hesitations at the time of the Orange Revolution in Ukraine. Vladimir Putin did well to lead Russia toward democracy. It's an imperfect democracy, but it's a democracy all the same. Yet this must not lead us to be complacent about behavior that is no less acceptable just because it takes place on the border between Europe and Asia. I was happy to see and admire German chancellor Angela Merkel's courage on these issues. From my point of view, France's image in the world and ability to get things done would benefit from a foreign policy that showed real intransigence on the issue of universal values. France has always incarnated these values, but it has not always defended them strongly enough.

THE MIDDLE EAST

We have a duty to work together to try and put an end to the conflicts that threaten to ignite the powder keg of the Middle East.

We cannot allow ourselves to remain impotent in the face of rising tensions and the aggressiveness of certain regional forces. And our recent experience shows us that when France and America work together we are effective.

This was the case with the passage of U.N. Security Council Resolution 1559 on Lebanon in September 2004. Unity between the United States and France made a cease-fire possible, helped lead to the departure of Syrian troops, and gave new hope to Lebanon.

More recently, I welcomed the August 2006 passage of U.N. Security Council Resolution 1701 and unreservedly supported Jacques Chirac's decision to send 2,000 soldiers to Lebanon to serve in the U.N. force there. The Israeli intervention in Lebanon may have been clumsy and disproportionate, but the truth is that there was just one aggressor in this conflict and that aggressor has a name: Hezbollah. Israel had the right—I would even say the duty—to defend itself and its citizens. It was Hezbollah that decided to take the Lebanese people hostage in a senseless adventure. Let's be clear: it is in Israel's interest to act proportionately. Even though it is the victim, it must do everything it can to avoid seeming like an aggressor.

We must also deal with the question of the Iranian nuclear program. France's position is clear and unambiguous: under the nuclear nonproliferation treaty, Iran has the sovereign right to acquire civilian nuclear capability—but it does not have the right to acquire a nuclear weapon. Yet everybody knows that there are too many suspicions about Tehran's real intentions. This is a risk we cannot take.

Through its support for Hezbollah and through its president's unacceptable remarks on the Holocaust and the existence of Israel, the Iranian regime has made itself an outlaw nation. The prospect of such a regime armed with weapons as destructive as nuclear missiles is terrifying. It would open the way to a murderous arms race in the region, as other countries would also want to pursue nuclear weapons. A nuclear Iran would also be a constant threat to Israel's existence. History has demonstrated the consequences of complacency in the face of aggression and fundamentalism. Resolving this question will demand the utmost firmness and the greatest unity. Diplomacy must be our main weapon, but I believe we must leave all options open in order for diplomacy to work.

The debate on the Iranian nuclear program must prompt the international community to take strong action on the global nuclear energy market. I am convinced that nuclear energy will continue to represent a major solution for the future when we are faced with a shortage of fossil fuel. So why not create a "World Bank" for nuclear fuel under the auspices of the International Atomic Energy Agency (IAEA)? The nuclear nations would contribute to it financially or in kind, and it would guarantee shipments of civilian nuclear fuel as well as the reprocessing of fissile materials to all nations that desire to develop nuclear energy for peaceful purposes while naturally renouncing the military nuclear option. In this way, the international community, represented by the IAEA, could offer all the guarantees of secure access to the benefits of nuclear energy without the risk of its being hijacked for military ends. Moreover, this would deprive certain potential proliferating nations of the pretext—and that's what it is—that they have the right to civilian nuclear energy and energy independence. This would help ensure that civilian nuclear programs would not be hijacked for nuclear purposes.

Finally, it is necessary for us to speak to the billion Muslims around the world. France has a role to play and a voice that is listened to in the Arab and Muslim worlds. But we must not forget that this world is not uniform. There are many Muslim and Arab worlds. The very concept of an "Arab policy" is nonsense. We have to develop and put in place a policy tailored to each of the regions of this world and not allow ourselves to be blinded by a unity that is only apparent.

We cannot make our relations with Israel conditional on the ups and downs of our interests in Arab societies. I feel close to Israel. Israel is the product of the Holocaust, which is a stain on the twentieth century and all of human history. All democracies are accountable for Israel's security, which is nonnegotiable. That in no way prevents France from expressing disagreements with the

Israeli government. But these disagreements, however major they might be, cannot call into question our relations with this small but so symbolic country, whose democratic practices and economic performance can only be admired. At the same time, we have to affirm the nonnegotiable right of Palestinians to have an independent state.

THE GREAT INTERNATIONAL DEBATES

France must finally play an active part, in international organizations, in all the great debates that our planet depends on. It must engage on the environment, which means absolutely respecting the Kyoto Protocol on greenhouse gas emissions and involving the United States, India, and China in emissions reductions. It must participate in the debate about massive increases in development aid, which are indispensable to world stability. It must support global mobilization to stop pandemics, beginning with AIDS. It has to help deal with nuclear proliferation issues, of primary concern to the nuclear power that France is and must remain.

There is no lack of subjects and this list is certainly not exhaustive. To play a greater role in these debates, we have both to set a better example at home and increase our strategic presence in negotiations and international organizations. Being stronger doesn't mean being tougher; it means being more indispensable. But our current policy of linguistic intransigence renders us inaudible. In the name of defending the French language, we refuse to speak other languages in international negotiations, including the informal discussions that are often most important. We thus manage to be simultaneously perceived as arrogant and excluded from the debate! At the same time, we have cut by thousands the number of places for French teachers abroad as well as scholarships to allow foreigners to study French in France. This is completely contradictory. I think it's time we got beyond this hypocrisy, of which we are the primary victims. Of course, we should continue to demand that

French be spoken when the official status of our language requires it. But we should also develop our ability to understand and speak other languages, especially English, fluently and even in discussing technical subjects. It's only in this way that we'll get our partners, in turn, to understand that they have an interest in speaking French.

Among the international issues, the reform of the U.N. Security Council is one that has to be dealt with. I am convinced that we can no longer limit ourselves to the current permanent members. Who really thinks that when it comes to ensuring world stability we can leave out countries such as India, Brazil, Japan, South Africa, and of course Germany? We have every interest in calling for and promoting this reform. That will spare us from having to be subjected to it, since I think it's inevitable.

THE NECESSARY DEBATE ON GLOBALIZATION

We need to have a debate about globalization. But don't count on me to be a naysayer. Yes, we have to give ourselves the means to deal with globalization as best we can. But let's not denounce globalization. It's a reality. We have to see it for what it is.

I am working for humane globalization, that is, globalization that promotes human emancipation and progress and rejects subjugation. Globalization is a unique opportunity to spread respect for human rights and democracy, to make knowledge available to all, and to allow millions of men and women to have access to the fruits of economic development. We too often forget all this. On the other hand, globalization should be condemned when it leads to child labor or men and women working unbearable hours under horrible conditions for a pittance and without any social benefits. It should be condemned when it drives thousands of people into illegal immigration. It should be condemned when it leads to a brain drain. It should be condemned when it neglects environmental concerns and forgets that the price of a good is not only what it cost to make it, but also the cost of the environmental damage that results from its production and transportation. I want trade negotiations at the WTO to take into account environmental issues as well as the issue of social protections in countries we compete with.

But to convince our partners of the interest, necessity, and feasibility of humane globalization—which would be a fine grand design for France—we have to go along with globalization. France must accept the best to fight the worst. It cannot act like the Gallic village surrounded by Roman camps, forgetting that only in Asterix cartoons does the Gallic village win.

THE ENVIRONMENT

We cannot ignore the growing threat to our environment. We must save the planet from the consequences of human activity. Half of our old-growth forests have already disappeared from the earth's surface; the ice is melting in Greenland; world carbon dioxide emissions will increase by at least 75 percent in the next 25 years, and the situation is already critical. No one would be spared the consequences of worldwide climate disaster.

The greenhouse effect, ocean pollution, the looting of natural resources—in thirty years there may be no more oil, in a century no more gas—how will we face these energy challenges? And this looting of natural resources will not just mean climate change, desertification, a loss of biodiversity, the degradation of health, and the endangerment of the most fragile forms of life. It could also lead to wars of hunger and wars over water. And these wars will be all the more terrible because they will be wars of desperation. Those waging them will have nothing to lose, because they will have nothing left.

France and the current U.S. administration have different views on this issue. But as I've said, friends can have disagreements. We should talk about them calmly, without pitting various parties against one another.

The problem of the environment isn't only a quality-of-life

problem but a problem of life itself. We need the United States to wake up and make this battle for the protection of our planet a battle of American democracy. How could it run counter to the values America defends? Americans respect life and have placed life at the heart of the values of American democracy, so why and how could they be uninterested in the continuation of life on our planet?

GAULLISM TODAY

The French are nostalgic about the Gaullist era. This was a time when the French state meant something, France was respected around the world, France's representatives had a vision, and personal ambitions seemed to disappear before the only respectable collective ambition: the destiny of the country. The world has changed a lot since the time of General de Gaulle. Democracy plays a greater role, there is more freedom of communication and expression, and information circulates more quickly and widely. So politics is necessarily different, and France could not be led today as it was in de Gaulle's day. In foreign policy, the situation is also very different, in particular because of the development of the European Union in ways the general didn't support.

Still, I think Gaullism remains relevant—as a school of thought and an approach to politics—in a number of ways. I would even say that Gaullism has a particular relevance for France today, which in some ways is in a situation similar to 1958. Gaullism is first and foremost intellectual freedom. This freedom enabled a man who had been programmed in every way to save French Algeria instead to save France from the trap in which it was enclosing itself by refusing to end colonialism. Gaullism is also the choice of preserving eternal France by activism and reform rather than inertia. If you compare France's evolution over the past 30 years with the period following de Gaulle's reforms, you can easily see the respective merits of repetition and innovation. For French people at every level, Gaullism is a popular rallying point around love for the country and pride in being French. Finally, Gaullism is a certain idea of humanity. It's the belief that in every human being, there is a star that is shining, a secret dream that is waiting, and a hope waiting to be realized.

The saddest thing about France is the feeling of so many French people that their hopes will never be realized. The final clean break, the greatest and in my mind the most important one— the one that justifies all the others—is the break with this despair. We must give everyone new perspectives. We have to give everyone renewed hope. We must give people a new opportunity to realize their dreams.

A New France

The world has changed a lot over the past ten years. We've witnessed the September 11 terrorist attacks, the 2002 and 2007 French elections, France's rejection of the European Union Constitution, and the emergence of China and India. The age of abundance has ended, and an age of scarcity has begun.

France has changed. France is going through an unprecedented identity crisis. Its model of assimilation is failing, its social model is bankrupt, and its internal cohesion is disappearing. The French are beginning to doubt themselves. They doubt their values, their future, their identity, and their role.

I have also changed. Life has changed me. Now that I've passed fifty, I don't think about what my life means the same way I did when I was thirty or forty. I no longer feel the need to prove everything. I've become calmer and perhaps wiser too. I've gained perspective. I've learned from experience not to overact. Spending four years at the interior ministry was no doubt very important in all this.

I have changed because having power changed me. It showed me a side of human life that I was unfamiliar with. I saw real suffering, the great pain and loss of victims. I learned that deep down we all have inexhaustible energy if only we learn how to draw on it.

Unlike some others, I did not conclude from my experience

with power that politics is hopeless. On the contrary, having power convinced me of the enormous responsibility of politics in the difficulties that the French people face today. It convinced me that politicians can and must act. People sometimes call the attitude that nothing can get done the "culture of government." I want to see a different culture of government: the culture of responsibility, action, results, and the need and duty to act and get things done. I share Régis Debray's view that "a statesman is someone ready to face the consequences of getting what he wants." I feel ready. I'm ready to bear the consequences of my decisions and of the faith that the French have showed in me.

When you've seen misery and suffering up close and when you want to heed the desperate call of those who need you, you look at politics differently. It inevitably takes on a more serious dimension. Before I had power I underestimated this seriousness, but now I've taken it on board.

The heavy responsibility of power becomes clear when, as interior minister, you've had to confront the pain of parents of a girl burned alive by thugs, or when you've had to deal with riots that could kill innocent people at any moment. You can think of politics in terms of consequences and results only when, as finance minister, you have had the fate of thousands of threatened workers in your hands, or when you've had to face social exclusion, poverty, unemployment, and homelessness.

A politician feels responsible only for himself. A statesman knows that he's responsible for others and that he cannot escape this responsibility. He knows that his decisions could have major consequences for the lives of many people. I have come to understand how serious this is. You can govern well only if you feel directly concerned with those you are governing. You have to be ready for total commitment. I have prepared myself for this. My political morality is that you do not have the right to ask others to sacrifice if you're not prepared to sacrifice yourself.

I'm no longer a political warrior. I seek no longer to conquer, but to convince. I have adversaries, but no enemies. I do not think about revenge. The concept of hatred is alien to me. I am not nostalgic. I'm always looking forward. I need to do things and build things. I have set a clear goal: to make the battle against injustice the warhorse of the republican right, because if I'm to fight anything it is injustice, inequality, and violence. I have always believed that the lack of clear choices is democracy's worst enemy. That's what weakens democracy. Democracy dies away when there is no longer any difference between the majority and the opposition, when the left and the right are no longer faithful to their values, and when no one is willing to stand and fight for the policies for which he or she was elected. This is no doubt one of the main causes of the current crisis of politics.

BUILDING FRANCE

My focus as president will be on what I want to build. François Mitterrand and Jacques Chirac are statesmen who focused more on history and on French traditions than on reforming France. It was their right to do so, after all, and it was the right of the French people to elect them. I think my energy and enthusiasm are better suited to be put to use for renewal. This may be destiny: the destiny of an encounter between a man and a woman at a particular moment in their lives, and a country in a particular condition at the very same moment. What interests me is the modernization of France. I imagine for my country only the top place in Europe. I want a future for each French citizen. I want each one to have a right to social advancement and I want each person's merit to be recognized. I think the French people are waiting for the arrival of the France of the future. They want, not a radically different France, but a France in which today's obstacles are removed so that everyone can realize his or her hopes, dreams, and ideals.

This new France should first and foremost be a free country, a country where things can be said without incurring disapproval, and where creative thinking is valued. It should be a country where there is no discrimination on the basis of one's skin color, one's name, or the neighborhood in which one lives but where the right to succeed is guaranteed to all who do what it takes. In this country,

individuals will be entitled to hold their beliefs and to practice religion without being labeled bigots or terrorists. They will be able
to put their children in the school of their choice. People will be
able to enroll in school at any time, regardless of the subject they
specialized in, or the career track they embarked on at age fifteen.
Creating a business will be possible, and even encouraged and valued, rather than being vilified or a cause for suspicion. In this country, individuals will be entitled to make their own decisions in life,
and their choices will be safeguarded more than their status. The
government will seek to resolve problems by providing a range of
policies, not by imposing a single course of action.

This new France will be an example of modern and responsible democracy. In this country, it will be possible to disagree while
maintaining respect for each other, and it will even be possible to
change one's mind. Debate within political parties will be seen not as
divisive, but as enriching. Ethical politics will replace underhanded
tactics. Parliament will be an effective counterweight to the executive. In this country, the justification for the government's existence
will be to act, not to survive. Politics will attract the best and the
brightest, and the people will gain a new sense of confidence in elites.
In turn, elites will not assume that people will accept reforms only
during the hundred-day period following a presidential election. In
this country, those who have power—whether they are involved in
public policy, justice, or business—will exercise power accountably.

In this new France, work, effort, and merit will pay off. Social
mobility will be a legitimate and realistic hope. Those who work
will always earn more than those who don't. Those who take risks
will always be better compensated than those who don't. Philan-
thropy will bring social recognition but not administrative red tape.
All will not be won or lost at age twenty-five depending on whether
you have the "right" or the "wrong" diplomas. It will be possible to
start over after having failed. In this country, success will be valued
because it is considered a common good. Being employed will be
enough to become a home owner. It will be possible to transfer to
one's children the fruit of one's labor without its being taxed away.

This new France will be capable of reconciling solidarity and responsibility. Its public schools will be the pillars of equal opportunity. This country will give more to those with greater handicaps. The sick will benefit from better treatment regardless of their social or economic status. The State will offer disbursements but will also offer human and personalized assistance to citizens until they can fend for themselves. This new France will lend a hand to those in a situation of need, but those in need will have to make the effort necessary to reach out and hold on to that hand.

This new France will be a country that stops producing only more debt and unemployment. It will strike a balance between its public finances and social accounts, by achieving growth and a return to full employment. It will once again be in a position to invest in the future. Losing one's job will no longer be a trauma, since finding a new one will be quick and easy. Companies will create innovative products, take over new markets, increase salaries, and add buying power.

This new France will be a country in which one's well-being increases thanks to access to sports, to more comfortable and dependable means of collective transportation, and to the spread of green spaces. There will be reductions in environmental pollution and efforts to make people's everyday lives easier, especially for women.

This new France will be a country that resumes its leadership of Europe and that is listened to in the international arena.

This new France will be a country reconciled with itself. Being French will again be equated with loving France, its eternal values, its exceptional destiny, and its universal culture. This will be a France in which the expression *français de souche*—"native French"—has disappeared, in which diversity is seen as a source of enrichment, in which everyone accepts and respects everyone else's identity. Fights over who is "more French" will give way to an equality among citizens that will finally become reality.

I have called this the France of the future. In fact, it is a France that has always existed but has not been apparent in recent years.

I think there are many of us out there dreaming of such a France and wanting to create it. I believe in France's destiny. I imagine for France a future equal to its past.

Nicolas Sarkozy was born in 1955 in Paris. Trained as a lawyer, he has served as president of France's governing political party, the right-of-center Union pour un Mouvement Populaire (UMP, Union for a Popular Movement), mayor of Neuilly-sur-Seine, and, in earlier national governments, as minister of the budget; minister of communications; minister of the economy, finance, and industry; and as minister of the interior. He was elected president of France in May 2007.

Philip H. Gordon is Senior Fellow at the Brookings Institution. He has written about international affairs for the *New York Times*, the *Washington Post*, the *Wall Street Journal*, and *Foreign Affairs*. His latest book is *Winning the Right War: The Path to Security for America and the World*.